Early Farm Tractors

A History in Advertising Line Art

This image is from an "International Gasoline Traction Engine" advertisement in *Implement & Vehicle Journal*, Oct. 8, 1909. It is an example of International Harvester's Type A Tractor, built from 1907-1916.

Early Farm Tractors

A History in Advertising Line Art

Jim Harter

WingsPress

San Antonio, Texas
2012

First Edition

Print Edition ISBN: 978-1-60940-252-5
ePub ISBN: 978-1-60940-253-2
Kindle ISBN: 978-1-60940-254-9
Library PDF ISBN: 978-1-60940-255-6

Wings Press
627 E. Guenther
San Antonio, Texas 78210
Phone/fax: (210) 271-7805

On-line catalogue and ordering:
www.wingspress.com
All Wings Press titles are distributed to the trade by
Independent Publishers Group
www.ipgbook.com

Library of Congress Cataloging-in-Publication Data:

Harter, Jim.
    Early farm tractors : a history in advertising line art / Jim Harter. -- First Edition.
        pages cm
    ISBN 978-1-60940-252-5 (hardback : alk. paper) -- ISBN (invalid) 978-1-60940-253-2 (epub ebook)
-- ISBN (invalid) 978-1-60940-254-9 (kindle ebook) -- ISBN (invalid) 978-1-60940-255-6 (library pdf
ebook)
    1. Farm tractors in art. 2. Commercial art--United States--History--20th century. I. Title.
    NC1002.F37H37 2012
    740'.496292252--dc23
                                          2012022587

ACKNOWLEDGEMENTS

    I first wish to acknowledge Bryce Milligan of Wings Press for having faith in this project and being
willing to publish it. The majority of these images came from copies made at the Texas A&M University
Evans Library at College Station. This fine library has a wealth of agricultural periodicals from the early
twentieth century. I wish to thank the library for my well spent time there. A few images also came from
the Research Center of the Panhandle-Plains Historical Museum in Canyon, Texas. I would like to thank
the Director, Warren Stricker for his help. I also am grateful to Canon Copier Co., which unlike some of
its competitors, produces machines capable of making fine quality reproductions of highly detailed drawings
and wood engravings. During several lengthy visits to College Station I traveled fifty miles each evening to
stay the night with my relatives Kay and Carl Evans, owners of Texas Pneumatic Tool Co. of Reagan, Texas.
I wish to thank them for being gracious hosts. Finally, I am grateful to family members and friends Ralph
& Bennett Kerr, Mike Harter & Monique Dupuis, John & Marty Marmaduke, Jim & Barbara Whitton, Ed
Conroy, Marjorie Robbins, Eric Edelman, James Hendricks, Joan Hall, Nathan Sumar, and Terrelita Maver-
ick for their continued support.

Contents

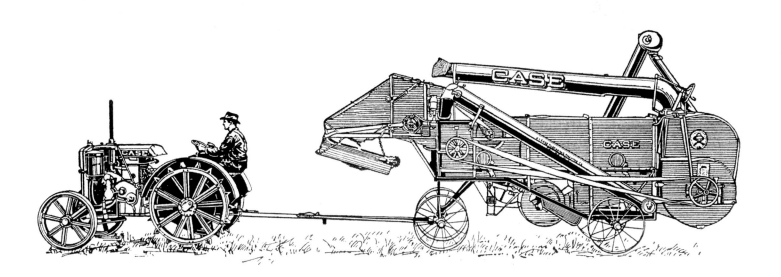

Dedication

Born in 1941, I grew up in Lubbock, a regional Texas city and agricultural center. In the 1950's, as a teenager, I received an intimate introduction to tractors through my relatives, the Joe Erickson family of Perryton, in the Texas Panhandle. They arranged for me to work for the Cecil Leicht family of Ochiltree County. For two wonderful summers I learned the value of hard work and country living. In the aftermath of the wheat harvest I plowed day and night. Otherwise I helped in baling and stacking hay, digging postholes, repairing fences, and performing various other farm chores. In appreciation for this experience, this book is dedicated to the memory of my relatives Joe and Annabeth Erickson and their children Charlie, Ellen, and John. It is also dedicated to the memory of Cecil and Margaret Mary Leicht and their children Johnny, Mary Cecilia, Jeff and Bert.

MANY BOOKS have been published on tractor history over the years. Some have been general histories, while others have focused on single companies like John Deere, International Harvester, Caterpillar, Case, or Oliver. A rarer type, like C.H. Wendel's well-known *Standard Catalog of Farm Tractors*, provides encyclopedic information on virtually every American tractor company that existed. This present volume is yet another kind, one where art and aesthetics have emphasis. It is a unique presentation of advertising line illustrations from the early period of tractor manufacture through 1929. Besides having graphic design value, these pictures evoke the spirit of that time and provide an interesting glimpse of the constant engineering and design changes that unfolded as this new technology evolved.

The reader is invited to enjoy these images on their own terms, but to also consider the larger aesthetic mystery, of how form follows function, and of how something as ordinary and utilitarian as a tractor seems to have inherent standards of good design, correct proportion, and beauty. At the same time we all know that beauty is elusive and multifaceted. Perhaps ultimately these standards exist only as a Platonic ideal. An ideal that a good industrial designer can try to intuit, and in so far as he is able, bring into our material world. Compiled and written by an artist who has edited many collections of historic clip-art, this book is intended for tractor enthusiasts, historians, artists, illustrators, students of industrial design, and lovers of graphic art.

Illustrations in tractor advertisements became common about 1912. Because of cost, however, very few were in color. Most illustrations were not of high quality, and the great majority consisted of photographs showing tractors in a work situation or in a side view. Line drawings were more expensive than photographs, but they produced a cleaner and bolder image on the page. This was desirable when low quality paper was used, as was normally the case with farm journals. Line drawings were done in different styles and varied in level of detail.

The tractor images selected for this book come from agricultural journals like *Farm and Ranch*, *Breeder's Gazette*, *Progressive Farmer*, *Country Gentleman*, *California Cultivator*, *Farm Implement News*, *Wallaces' Farmer*, and Britain's *Mark Lane Express*. Organized chronologically as much as possible, they provide pictorial documentation of individual tractor models; the design evolution taking place both within companies and overall; and illustration styles of the period. Here one will find early models from surviving companies like John Deere and Caterpillar. Also are popular brands of the time such as Rumely, Hart-Parr, and Wallis, plus many obscure makes like Fageol, Nelson, and Thorobred. While this book focuses on American tractors, it includes a few examples of British and Canadian brands.

Tractors, automobiles, and trucks made their historic debut about the same time. This was because all were powered by the newly developed internal combustion engine. Pioneered in the 1890s, tractors emerged during a period when steam traction engines were used for plowing and other farm purposes. In the early 1900s standard production began on a few tractor models. By 1920 this technology had surpassed steam, and was rapidly accelerating the pace of farm mechanization. An amazing change in farming methods and production took place as a result. The high costs of steam traction had largely limited its use to large-scale grain production, and those areas where it was grown. Otherwise, horses and mules remained the primary source of power for small and average sized farms. All this was to change as tractors evolved from ponderous converted traction engines to smaller, lighter and more versatile machines.

During the First World War this mechanization process was expedited as men, draft animals, and fodder were diverted from the farming sector just as the need for agricultural production intensified. Beginning at this time many new companies entered the tractor market. This served wartime and immediate postwar needs, but beginning in 1920 overproduction of crops occurred and prices collapsed. One of the new entries in 1918 was Henry Ford's Fordson Model F, the first mass-produced tractor. At this time Ford's Model T automobile enjoyed great popularity with farmers because of its reliability, simplicity, and cheap price. This reputation helped the Fordson quickly gain a large share of the tractor market, and by 1922 it sold a greater number of tractors than the entire competition combined. Fordson's main competitor, International Harvester, fought back and in 1924 introduced its McCormick-Deering Farmall, a revolutionary row-crop tractor that eliminated any further need for draft animals. By the end of the decade International Harvester had returned to dominance and Fordson had dropped all U.S. production.

While America had only 600 tractors in 1907, there were over 500,000 by 1925. However, tractor demand in the 1920s was reduced because of depressed farm product prices. This intensified competition among tractor manufacturers, forcing many out of business. In 1921 there were 186 firms listed, but by 1929 only 47. However the export market was good. In 1926 approximately 50,000 wheel style tractors and 1,200 crawlers were exported abroad. These products went to countries like Canada, Australia, Russia, France, Italy, and Argentina. America's remaining companies had a peak year in 1929, producing 229,000 tractors. However, a precipitous decline followed as the Great Depression began. This reached a nadir in 1932 when only 19,000 tractors were built. By this time though, farm mechanization had largely prevailed, substantially improving productivity and reducing labor needs. This resulted, during the 1930s, in a mass exodus of workers from the countryside into the city.

A tractor history isn't complete without mentioning the importance of the previous steam traction era. The reader is provided with an overview of this at the beginning of the text. This is supplemented at the beginning of the pictorial section by six wood-engraved plates that show evolutionary phases of this technology. These pictures came from publications like *The Engineer*, *Scientific American*, *American Machinist*, and the J.I. Case Threshing Machine Company catalog of 1899. Following these is a seventh plate, showing an English kerosene-powered traction engine of 1896. It provides an example of the transition phase from the old technology to the new. Forty-eight more plates follow, gathered from tractor advertising images.

In searching for tractor pictures, I began by looking through various farm journals from the first decade of the twentieth century. However, there were very few tractor illustrations, and with one exception, pictorial quality was poor. However, beginning in

1912 many tractor advertisements began to have a more professional look. Thus, aside from one image from 1909, all depictions in this book come from tractor advertisements of 1912 or later. Luckily a few of these showed tractor models already in production for several years. Tractor models were often manufactured for a number of years before they were significantly modified or discontinued. When this information is available the production years are indicated rather than the specific year of the advertisement.

While the steam traction images are virtually all wood engravings, only a few of the tractor pictures are. Two examples can be found on Plate 13, the Leader Tractor and the Hackney Auto Plow. This technique, very popular in the nineteenth century, was good for highly detailed work, but it was relatively expensive and time consuming. It largely went out of fashion once half-tone photographic reproduction was perfected in the early 1890s. However, limited use of wood engravings continued into the 1920s for advertising and catalog illustration. In this process boxwood was sawed crossways into slices about 3/4" thick, then the height of type. Upon a single block a sketch was made in reverse, and then with various tools an engraving was fashioned. Once the block was set with type, printing could begin.

Many images here were probably done on scratchboard, a later technique that could create fairly detailed drawings reminiscent of wood engravings. This involved using a piece of cardboard stock coated with impermeable white clay. Upon this surface areas of India ink could be applied where needed. Various tools were then used to scratch or scrape away patterns on the inked surface, revealing the white below and creating a suitable picture. Other images appear to be pen and ink drawings, rendered in various styles. In preparing this book I have included the best line style images I was able to find. The traction engine pictures have been scanned at 1200 dpi and the tractor images at 600 dpi. All have been touched up to improve graphic quality. These have been printed in as large a format as possible to pass them on virtually intact to future generations. It is my hope that you the reader will enjoy this book as much as I have had in creating it.

Wood engraving from 1885 catalog of Springfield Engine & Thresher Co.
showing steam traction engine powering a swinging stacker.

Early Farm Tractors

Steam Ancestors: Traction Engines & Cable Plowing

THE TRACTORS WE SEE PLOWING in fields today are the result of a long process of technological evolution. This evolution began with the idea of developing a machine to substitute for the animal power used in plowing. Whether or not they were the first to entertain this notion, David Ramsey and Thomas Wildgosse were granted a patent in England on January 17, 1618 for "Newe, apte, or compendous formes or kinde of engines or instrumente… to ploughe grounde without horse or oxen, and to enrich and make better and more fertil as well…." While nothing resulted from this vaguely worded patent, it nonetheless indicated that human minds had begun dreams that continue to unfold today in each new advance of tractor technology.

In the 1760s Englishman James Watt developed a workable steam engine. By the 1830s this new technology had reached a level of refinement where its application to plowing seemed inevitable. Robert Stephenson had invented the multi-tubular boiler in 1827, the Liverpool & Manchester Railway had been inaugurated in 1830, and individuals like Oliver Evans and Goldsworth Gurney had created road machines. However there were many obstacles to overcome before *direct* steam plowing became practical. Field surfaces were more difficult for holding traction than hard roads and much of the power generated by heavy inefficient engines would be expended just to move them, leaving little extra for the drawbar.

One early pioneer of direct plowing was the Englishman J. Boydell. In 1846 he received a patent for an "endless railway" system whereby flat wooden "shoes" or rails were attached to wheels supporting a portable engine. By 1857 a steam powered traction engine capable of both plowing and road haulage had been built to his specifications. The designation "traction engine" indicated that it was a self-propelled machine; apparently in this case it required horses for steering. Boydell's "Steam Horse," built by Messrs. Charles Burrell of Thetford, was successfully demonstrated at Croxton and Louth. It had five wooden shoes on each of its two drive-wheels and two guide-wheels. However it couldn't be operated at speeds over 2½ mph or the wooden shoes would break off. A small number of other Boydell engines were subsequently built.

On July 20, 1859 J.W. Fawkes of Lancaster, Pennsylvania, successfully demonstrated his traction engine at Philadelphia. Having an upright boiler and two cylinders, it pulled a gang of eight prairie plows that could be raised or lowered as need be. The single drive wheel consisted of an iron drum six feet in diameter, and six feet wide, through which an axle extended. In a related development that year, Englishman Thomas Aveling converted a Clayton & Shuttleworth portable steam engine into a traction engine. Years later his Aveling & Porter Co. of Rochester would become one of the earliest suppliers of traction engines for direct plowing. However at this time, a lot more work was required before this technology became practical.

In the meantime an alternative system emerged, known as steam cable plowing, that was to compete with direct plowing for a number of decades. In its earliest form it involved a portable steam engine, equipped with a winding spool, on one side of a field, and an anchor cart with a pulley at the opposite side. In between was run a cable, creating an "endless rope" to which was attached a plow. The plow was pulled between the engine and anchor cart. As rows were completed both engine and anchor cart were pulled forward by horses on parallel paths. The need for horses was later eliminated as traction engines replaced portable engines for winding purposes. Early cable plowing pioneers included John Fowler of Leeds who developed a two-engine system, and a Mr. Hannam of Berkshire.

Hannam was the first to experiment with the "Roundabout" system. In this technique the winding engine was placed in one corner of the field and three portable anchor carts positioned in the other corners. While others experimented with this technology, Fowler's company was most successful in developing and marketing it. The company pioneered its own cable plowing "tackle" in 1852, patented its two-engine idea in 1856, and first demonstrated double engine plowing in 1863. Fowler was also credited with developing the balance plow. Even as late as the Paris Exhibition of 1878, cable plowing remained dominant over direct plowing. However in the United States only a few attempts were made to use this technology, and it never caught on because of the expense. Once this technology became obsolete it must have had its adherents. A friend of the author described seeing its use in France as late as the early 1970s. Perhaps the main advantage of cable plowing was that it kept the heavy weight of the steam engine off the plowing area.

During the 1870s a long effort of experiments in direct plowing began to pay off. Steering mechanisms improved, engine efficiency increased, and other obstacles were overcome. Practical traction engines emerged in Britain during this period, but they were large, heavy, and expensive. Besides plowing they could pull wagons on roads, and in a stationary position, using pulleys and belt, run threshers, saws, pumps, grinders, etc.

In the depressed American economy of the 1870s, however, British style plowing was unaffordable. In the early 1880s smaller portable engines began to be converted into traction units. Soon a number of companies were offering fairly reliable machines affordable for large-scale farming. These engines were quickly put to work breaking much of the prairie of the Midwest and West into productive farmland. Incremental improvements continued to be made in this technology until it was discontinued in the 1920s.

Among the many American manufacturers of steam traction engines were the J.I. Case Threshing Machine Co., Racine, Wisconsin; Frick & Co., Waynesboro, Pennsylvania; Geiser Mfg. Co., Waynesboro, Pennsylvania; Aultman & Taylor Machine Co., Mansfield, Ohio; Pitts Agricultural Works, Buffalo, New York; A.B. Farquhar Co., York, Pennsylvania; and Nichols & Shepard Co., Battle Creek, Michigan. British manufacturers of traction engines and cable plowing tackle include Robey & Co., Lincoln; Ransomes, Sims & Jefferies, Ipswich; Charles Burrell & Sons, Thetford; R. Hornsby & Co., Grantham; and Clayton & Shuttleworth Co., Lincoln.

Although this technology adequately served large acreage grain farms in America, Canada, and elsewhere, it was never well suited for average sized or smaller farms. Not only heavy and expensive, steam traction engines had a wide turning radius and were not very maneuverable. Thus it was easy for them to roll over in rough terrain, or get mired down. Traction engines burned coal, wood, or straw, and there was a danger of fires starting from their sparks. They were complicated, required constant maintenance, and on occasion could explode. They had to be fired up in the early morning to build steam pressure, and extra crewmen were required for bringing in the necessary quantities of fuel and water. Their heavy weight could damage soil structure and while virgin prairie sod supported them in the initial plowing, it was not always the case afterwards. In such instances farmers had to return to the use of draft animals.

Steam power is most efficient where few restrictions apply to engine size and weight; thus it is ideal for use on ships or in power plants. While it worked well enough for railroad locomotives, attempts to build smaller sized engines to incorporate into rail cars and trams proved problematic. These units had high maintenance costs and short longevity. For the same reason the application of steam technology to farm power reached its limits in the large and ponderous traction engine. Because of prohibitive cost they were normally only purchased by large-scale grain farmers or custom threshermen. Everyone else used animal power. What most farmers needed, although it probably wasn't fully understood at the time, was a cheaper, smaller, lighter, and more maneuverable traction engine. However, this would require an entirely new technology as well as a lengthy developmental period.

Early Developments in Gas Traction

The one technology offering this possibility was that of the petroleum powered internal combustion engine. In the 1890s the internal combustion engine was still in its infancy. However, pioneering efforts go all the way back to John Barber who built a crude motor fueled by coal gas in 1790, and Robert Street who built a turpentine-powered engine in 1794. Circa 1850 Etienne Lenoir began production in France of a gas-powered engine. It should be noted here that this was not powered by gasoline but rather by some form of vaporous gas. In 1862 Alphonse Beau de Roches conceived the idea of the four-stroke engine.

During the 1860s the Germans Nikolaus August Otto and Eugen Langen formed a company to build an improved version of Lenoir's motor. After winning a gold medal at Paris' World Exhibition in 1867, they continued their efforts, and in 1876 introduced the much quieter four-stroke Otto gas engine. In his book *The Illustrated History of Tractors*, Robert Moorhouse notes that this new motor introduced the "basic engine cycle of operation used today… 1. Induction of the air and fuel mixture, 2. Compression of the mixture, 3. Power from the burning of the fuel, 4. Exhaustion of the burnt gases and so back to induction."

Otto and Langen's firm became known as Gasmotoren Fabrik Deutz AG in 1880. Working for the company at this time were Wilhelm Mayback and Gotlieb Daimler. This pair subsequently left the firm and began engine experiments using a liquid petroleum derivative called benzin (gasoline). They developed a 900 rpm engine in 1885 and used it to power a motorized bicycle. About this same time Karl Benz, a German builder of gas motors, built a four-stroke engine and adapted it to power a tricycle. Earlier, in 1877, James Starley, a pioneer cyclist devised the first differential. A decade or so later Rudolph Ackerman conceived a geometrically correct steering system. All of these developments were important contributions to what would ultimately result in among other things, the farm tractor.

In 1886 the Otto engine was introduced to America by Philadelphia's Schleicher, Schumm & Co. who began manufacturing it under license. One person inspired by this new technology was John Charter of Stirling, Illinois. He began building gasoline-powered engines, taking out patents for each new innovation. In 1889 Charter mounted a single-cylinder 25-hp motor on the chassis of a Rumely steam traction engine. Sold to South Dakota farmer L.F. Burger, it provided belt power for a threshing machine. This creation was successful enough to result in the sale of six additional hybrid machines to other South Dakota farmers. However, no others were made.

In 1892, John Froelich mounted a 16-hp Van Duzen one-cylinder gasoline motor on the chassis of a Robinson & Co. steam traction engine. For this machine he also devised his own transmission and steering system. Froelich's machine was demonstrated successfully before a crowd of farmers at Langford, South Dakota, where it threshed 62,000 bushels of wheat in fifty days without a breakdown. In 1893 Froelich and a number of Waterloo, Iowa, businessmen formed the Waterloo Gasoline Traction Engine Co. However, Froelich's machine, despite its impressive performance, didn't catch on. In order to survive, the company began building stationary gas engines. Only in 1912 did it reenter what by then had become the tractor business.

In 1892 America's largest manufacturer of steam traction engines, the J.I. Case Threshing Machine Co., tested a gas traction engine designed by William Paterson. His strangely designed motor, which had a single cylinder and two pistons, proved unsuccessful, and Case dropped the project. In 1893 the Van Duzen Co., maker of the engine used by Froelich, tested its own gas traction engine. These design rights were subsequently sold to Huber Manufacturing Co. of Marion, Ohio, a steam traction engine builder. While Huber introduced a one-cylinder gasoline traction engine in 1898, and built about thirty afterwards, it didn't receive enough return on its investment to continue production. However, like Waterloo, Huber also returned to the business later on. Schleicher, Schumm & Co., which changed its name to Otto Gas Engine Co., entered the gas traction engine business in 1894. By 1900 the company had built only 14 machines. However limited production by the company continued until 1914.

In 1897 Kinnard & Sons Manufacturing Co. of Minneapolis began production of the "Flour City" gas traction engine. It sold four in 1898 and built 28 in 1899. The first gas traction engines considered really successful were those built by Hart-Parr Co. of Charles City, Iowa, in 1902-03. Charles W. Hart and Charles H. Parr were two engineering students who met at the University of Wisconsin in 1892. Sharing a mutual enthusiasm for internal combustion motors, they started the Hart-Parr Gasoline Engine Co. at Madison, Wisconsin, in 1897, initially building stationary engines. In 1901 they moved the operation to Hart's hometown of Charles City, and built their first gas traction engine prototype in 1902.

The following year fifteen machines were manufactured in two models; *17-30* and *22-40*. About this time horsepower began to be indicated by two numbers. The first indicated drawbar horsepower available for plowing. The second showed belt-pulley horsepower available for running machinery from a stationary position. In overall performance, Hart-Parr's early machines showed little improvement over contemporary steam-powered models. However, the firm's oil cooled two-cylinder valve in head motor gave good performance. The fact that five of the original 15 machines remained in service in 1930 attests to their durability. The early engines were apparently all used for threshing, but in 1906 some were tried for plowing. In 1907 Hart-Parr decided to differentiate its product from steam traction competitors by renaming it "tractor." This term subsequently came into general use. However, in his book *The Agricultural Tractor: 1855-1950*, R.B. Gray has noted that the word first appeared on a patent granted to Geo. H. Edwards of Chicago in 1890.

Early British Tractors

In his book, *The Illustrated History of Tractors*, Robert Moorhouse notes that the word tractor has its roots in the latin word *tractorius* which described "the act of drawing or pulling in a mechanical context." He also observes that this word was already starting to be used in association with petroleum powered traction engines, and he cites several examples. Among these was the British built Petter's Patent Agricultural Tractor exhibited in 1903. Elsewhere in his book Moorhouse mentions a *tracteur* built by the German Adolf Altman circa 1896. Altman was probably among the first in Europe to apply internal combustion technology for traction engine use. Unfortunately his machine was unsuccessful. An early British pioneer was Richard Hornsby & Co. of Grantham, Lincolnshire. It was among several firms building portable "oil engines." Licensed to build the Akroyd –Stuart kerosene engine, it used one in 1896 to power its Patent Safety Oil Traction Engine. Making its first sale in England, the company exported six to Australia in 1897.

In the late 1890s, John Scott of Duddingstone, formerly a Professor of Agriculture at the Royal Agricultural College, Cirencester, began designing a tractor. His first, called a "Motor Cultivator," was exhibited in 1900 at the Royal Show near Cardiff. Unlike converted traction engines, it was unique in design, and included a platform at the rear for carrying loads up to three tons. Besides having a belt pulley, it also featured a chain drive powered cultivator at the rear. A subsequent model, exhibited at the Royal Show near London in 1903, replaced the load-carrying platform with a seeding mechanism. In combination with the cultivator this allowed seedbed preparation and drilling in one operation. Scott was a visionary and he continued to explore revolutionary design ideas. Unfortunately his tractors, which were ahead of their time, achieved no commercial success.

More fortunate in this respect was Scott's contemporary, the engineer Daniel Albone, a racing cyclist and manufacturer of the Ivel Bicycle in Bedfordshire. Using his design skills he introduced the world's first practical lightweight tractor in 1902. Having three wheels the initial version had a mid-mounted 8-hp two-cylinder engine, single ratio forward and reverse, and cone clutch. Later models used more powerful Payne & Bates and Aster engines. Having only modest sales in Britain, the Ivel Tractor did much better in the export business, receiving many international awards. Unfortunately Albone died in 1906 at the age of forty-six. While his tractor remained in production until about 1916, few improvements were subsequently made, and soon it was no longer competitive.

James B. Petter & Sons of Yeovil, Somerset began experiments in the late 1890s initially producing a self-propelled portable machine for stationary work. The company's first tractor, called the "Intrepid," came out in 1903. The driver's seat was positioned above its one cylinder engine. It had a large belt pulley and a rear platform for hauling loads. Retiring from the business for a while, the company introduced its "Iron Horse" in 1915. This 3-wheel model was intended for pulling equipment previously drawn by horses. Its unique steering system employed the use of reins.

The most commercially successful of the early British tractors was the Saunderson built at Elstow, Bedfordshire. Having a 3-wheel design, the first of its "Universal" models won a silver medal at the Derby Royal Show in 1906. Its 30-hp engine, more powerful than its competitors, powered all three wheels. This tractor was the main component of a complete mechanization system. It could be used for stationary work, road hauling, or with the rear platform removed, used to pull a number of specially designed implements. In his book, *British Tractors for World Farming*, Michael Williams describes the versatility of this system. "….The demonstration started at 2 pm when the tractor began to cut a standing crop of wheat, using a binder. After a short time the binder was unhitched and the tractor pulled a threshing drum into the field. The thresher was unhitched, and the truck body was fitted and used to haul the sheaves of wheat to the thresher. Then the tractor pulley was used to drive the threshing drum, and later a grinder to make flour from the newly threshed wheat. Next, the tractor returned to the cleared stubble, and a plough was attached. When the stubble had been ploughed it was cultivated; then a seed drill was hitched to the tractor and a new crop of wheat was sown. Meanwhile the freshly ground flour was made into dough and a batch of loaves was baked. By 7 pm – five hours after the demonstration had begun – wheat had been harvested and turned into bread, and a new crop had been sown." Subsequent models of the Saunderson were also successful. Manufacture of the Saunderson was taken over by the Crossley company of Manchester in 1924. Other pioneer British tractors were built by Ransomes, Sims & Jefferies, Ipswich; Drake & Fletcher, Maidstone, Kent; Ideal Tractor Co., Birmingham; and Marshall of Gainsborough, Lincolnshire.

An Overview of Early American Tractor Development

In his *book Steam Power on the Farm*, Reynold M. Wik suggested that early American tractor development could be divided into three well-defined periods. These were: "(1) 1876-1902, Experimental state of tractor development. (2) 1902-1913, Manufacture of large gasoline and kerosene tractors. (3) 1913-1924, Development of the small gasoline tractor to the advent of the all-purpose row-crop tractor in 1924." My understanding of Wik is that his 1876-1902 period runs from the introduction of the four-stroke Otto engine to the creation of the successful Hart-Parr prototype.

Wik's 1902-1913 period describes a time when the first commercially successful tractors were built. These were largely modeled on contemporary steam traction engines and were designed to compete with them. Their fuel tank capacity usually ranged from 60 to 110 gallons. They had large drive-wheels, typically six to eight feet in diameter. These early tractors were often crude and frequently broke down. They had one or two-cylinder large-displacement engines, and operated at very slow speeds. A large flywheel was required for accumulating energy and distributing it evenly. Ignition systems were problematic, starting was often difficult, and dissonant sounds like backfiring were common. Primitive cooling systems were devised, sometimes using oil. Besides preventing winter freezing, oil systems allowed operation at higher engine temperatures, which worked better for low volatility kerosene fuel. Kerosene at this time was largely preferred over gasoline because its cost was cheaper.

On certain occasions tractor companies would demonstrate their products against competitors. In 1908 the annual Winnipeg Trials were begun and continued through 1913. In this important event both tractors and traction engines of many manufacturers were put through various tests in front of an audience of thousands of farmers, engineers, and representatives from many countries. Later on similar events were held in Fremont, Nebraska and elsewhere. In 1912 America's tractor industry was thriving, reaching a production total of 12,000 units. The market soon saturated, however, and for a couple of years afterwards the industry had to retrench.

Wik's 1913-1924 period runs concurrent with the development of smaller, lighter, and more versatile tractors culminating with the development of a successful row-crop tractor. At this time the old design paradigm was challenged and many new ideas were explored, although not all successfully. Among these were various three-wheel configurations, front-wheel drive and four-wheel drive. Another innovation was the *motor cultivator*, a tractor-like machine specially designed for row-crop use. However motor cultivators couldn't be used for plowing, so their impact was limited. Also at this time companies like Staude Mak-A-Tractor and Smith Form-A-Tractor sprang up, offering kits that converted automobiles like the Model T Ford into a tractor.

Important in upgrading the quality of American tractors was the Tractor Test Law passed by the State of Nebraska in 1919. This legislation, in response to poorly made tractors and false advertising claims, required tractors sold there to undergo standardized tests designed by engineers. These results were published nationwide, forcing America's tractor manufacturers to raise their standards.

Capable of running around the clock with increasingly less maintenance, tractors helped to reduce manpower needs. They also helped farmers eliminate the considerable costs associated with purchasing, feeding, and maintaining draft animals. Over the next several decades this freed up a vast acreage of productive land previously used for providing feed and fodder. This was no small thing. In his book *Wheels of Farm Progress*, Marvin Mckinley observed that: "Each growing season about five acres of productive land had to be allotted to every horse in the barn, for raising oats, fodder, hay and straw. Sustaining the 24 million horses and mules on American farms required a total acreage equal to the combined areas of Iowa, Illinois, Indiana and Ohio." However, teams of horses were still necessary for cultivating row crops, because tractors couldn't stay within rows. Finally in

1924, the first fully successful row-crop tractor was introduced, the IHC McCormick-Deering Farmall. Because of its special design, this one machine could both plow and work row-crops, and in the process it made draft animals obsolete.

Some American Tractor Manufacturers

ALLIS-CHALMERS

In 1901 the merger of E.P. Allis, a manufacturer of steam engines, and three other firms resulted in the formation of the Allis-Chalmers Manufacturing Co. of Milwaukee. It was mainly involved with the production of industrial machinery until 1914. That year the company introduced a lightweight three-wheel *10-18* tractor powered by a two-cylinder horizontally opposed engine. About 2,700 units were sold when production stopped in 1921. A more conventional design was the *18-30*, which had unit-frame construction. Its debut in 1919 was badly timed because of intense Fordson competition. However it was well built and after a rocky start it gained a good reputation. Around 1921 it was re-rated upwards to 20-35. It was replaced in 1930 with the *Model E 25-40*.

Built from 1919 to 1923 was the *Allis-Chalmers 6-12 General Purpose Farm Tractor*. Like its competitor, the Moline Universal Tractor, its drive wheels were in front. Various interchangeable implements could be attached at the rear. The *Allis-Chalmers 12-20*, powered by a four-cylinder valve-in-head Midwest engine, was built from 1921 to 1927. However it was re-rated after the Nebraska test, and became the *Model L 15-25* in 1922. Allis-Chalmers set up a separate Tractor Division in 1926. In 1928 it acquired the Monarch Tractor Corporation of Watertown, Wisconsin, a manufacturer of crawler tractors. In 1929 the company introduced the *Model U 19-30*. A very successful design, it remained in production until 1952. It was initially powered by a four-cylinder Continental engine. A row-crop version, the *Model UC*, came out in 1930. In 1932 the Model U became the world's first tractor to be fitted with pneumatic rubber tires.

AVERY

The Avery Co. of Peoria, Illinois was originally founded in 1874 in Galesburg, Illinois. Robert Avery, one of the firm's founders, had been an inmate of the infamous Confederate Andersonville Prison during America's Civil War. He spent his time well there, devising better ways to plant corn. This became the seed idea that inspired the company. By 1891 Avery had begun making threshers and steam traction engines. Later on the company began experiments with gas traction and in 1909 introduced a novel four-cylinder truck-tractor combination, the *Farm & City Tractor*. This model was built until 1914.

After failure of a large one-cylinder engine prototype in 1910, the company brought out its two-cylinder *20-35* the following year, having a horizontally opposed engine. This design was successful and in the next few years Avery introduced a number of other tractors intended to fill every market niche. Besides having horizontally opposed engines, they all had a singularly distinctive feature. In order to engage the proper gears, the entire engine and radiator were shifted forward or backward upon a sliding frame.

An improved version of the *20-35* was built from 1912 to 1915. Production of the *12-25* ran from 1912 to 1919. The *40-80*, powered by two combined two-cylinder engines, was sold between 1913 and 1920. The *8-16* was manufactured from 1914 to 1922. Also built in this same period was the *25-50*, powered by two combined 12-25 engines. The *18-36* used two combined 8-16 engines. The first tractor in the industry to have replaceable cylinder sleeves, it was sold from 1916 to 1921. The four-cylinder *14-28* was built between 1919 and 1924. The *45-65* was manufactured from 1920 to 1924.

Avery's first really lightweight model, the *5-10*, made its debut in 1916. In the first version the driver's seat was positioned ahead of the rear wheels. In the *5-10 Model B* it was moved to the rear. In 1919 the Model B was replaced by the lightweight six-cylinder *Model C*. An orchard version was also offered. Shortly later Avery introduced a six-cylinder *Motor Cultivator*, and about 1922, its *15-30* model. As other models became dated, cooling systems were changed and driver cabs removed. Despite these improvements, however, Avery went into bankruptcy in 1924, and was shortly reorganized as the Avery Power Machinery Co. Afterwards it paid little attention to tractors, focusing instead on more profitable aspects of its business. In the late 1930s, however, Avery developed a very fine tractor, the *Ro-Track*. Unfortunately World War II brought its production to a halt. It should be noted that there was another Avery firm, B.F. Avery & Sons Co. of Louisville. It produced a small number of tractors.

CASE

Having failed with its experimental tractor of 1892, the J.I. Case Threshing Machine Co. waited until 1911 before trying again. That year it introduced a heavy two-cylinder *30-60* model, which in 1911 won a Gold Medal at the important Winnipeg Contest. Its manufacture ended in 1916. 1912 saw the debut of the popular two-cylinder *20-40*. The early version of this large tractor was built until 1916 and a revised model until 1919. In 1913 it won two Gold Medals for fuel economy at Winnipeg. That same year Case completed its tractor works, and introduced a lightweight two-cylinder *12-25* model. It remained in production until 1918.

In 1915 Case brought out its three-wheel, four-cylinder *10-20*. Like some of its contemporary competitors, it featured a forward wheel for steering, a rear driving wheel, and a rear outrigger support wheel. It was advertised as late as 1920. A much more attractive design was the four-wheel, four-cylinder *9-18*, introduced in 1916. The engine of this popular machine was mounted crossways. Its production ended in 1918.

That year saw the debut of the first of a series of Case tractors having four-cylinder cross-mounted engines and unit-frame construction. The *10-18* was built until 1922, when it was replaced by the *12-20*. The 12-20 later became the *Model A*. It was a direct competitor with the Fordson Model F. The *15-27* was built from 1919 through 1924, being replaced by the *18-32* in 1925, which later became the *Model K*. The *22-40*, introduced in 1919, was replaced in 1924 with the *25-45*, which later became the *Model T*. The large *40-72* was manufactured from 1921 to 1923.

Production of the cross-motor series halted at the end of 1928 after Leon Clausen was installed as the new company president. That same year Case bought out an important competitor, the Emerson-Brantingham Co. of Rockford, Illinois. When Case bought the J.I. Case Plow Works name from Massey-Harris in 1928, the company name was changed to J.I. Case Co. For 1929 Case offered a new series of more conventionally designed four-cylinder tractors. These were the *Model L*; its downscale version, the *Model C*; and the *Model CC*, a row-crop version of the C. These tractors were well engineered and they carried the company safely through the Great Depression.

CATERPILLAR

The idea of crawler traction is old. An English patent was granted for this idea in 1770. One early steam crawler design, Parvin's Traction Engine, received a US patent on October 10, 1871. In 1890 a steam-powered crawler was built by Stockton Wheel Co. of Stockton, California. The name of the company was changed to Holt Manufacturing Co. in 1892. Benjamin Holt, owner of the firm, demonstrated a more successful version in 1904. Eight subsequent steam crawlers were built between 1906 and 1908. In 1908 Holt purchased the Daniel Best Co. This firm, also a builder of steam traction engines, had a little experience with internal combustion engines. In 1909 Holt acquired a defunct factory in Peoria, Illinois. Operations began there but also continued in Stockton.

In 1908 production began on gas-powered crawlers. In 1910 Holt registered the name *Caterpillar* with the U.S. Patent Office. The early Caterpillars were all of half-track design with a single wheel in front for steering. They had large four-cylinder engines with each cylinder cast separately. Offered in *45* and *60-hp* models, they became popular for large acreage farms. A smaller model, the *Caterpillar 18*, was introduced in 1914. It was very useful in orchards and vineyards. Another small model from that period was the *Baby Caterpillar 20-30*. In this design the front tiller wheel was eliminated. The *Junior* replaced it in 1915.

Another model eliminating the tiller wheel was the *25-45*, offered from 1916 to 1919. It had two forward speeds and weighed 13,900 pounds. Still retaining the tiller wheel were the *50-75*, built from 1916 to 1919, and the *70-120*, built from 1918 to 1921. During the First World War Holt Caterpillars were used by both sides for pulling heavy artillery. The British converted Caterpillars into the first war tanks. Holt introduced the *10-Ton* and *5-Ton* Caterpillar models in 1919. These more modern designs eliminated the front tiller wheel for good. In 1923 Holt introduced a *2-Ton* Caterpillar.

From about 1910 onwards a number of other firms experimented with crawler designs and some went into production. Among Holt's competitors were Yuba Manufacturing Co. of Maryville, California; Cleveland Tractor Co. of Cleveland, Ohio; Monarch Tractor Co. of Springfield, Illinois; and C.L. Best Gas Traction Co. of San Leandro, California. C.L. Best was the son of Daniel Best, whom Holt had previously bought out. Becoming one of Holt's strongest competitors, he developed a very similar line of tractors called *Tracklayers*.

In 1925 Holt and Best merged to become Caterpillar Tractor Co. of Peoria, Illinois. Under the new company production of some models continued while others were terminated. The Caterpillar *Thirty* and *Sixty* models were retained from the Best line. New Caterpillar models included a *2-Ton*, built from 1925 to 1928, and *5-Ton* and *10-Ton* designs sold only in 1925 and 1926. These machines were all equipped with selective sliding gear transmissions. From 1928 to 1932 the *Caterpillar Twenty* was offered. Between 1929 and 1931 the *Ten* and *Fifteen* models were built. In 1929 Caterpillar secretly turned its attention to developing a diesel engine. Perfecting a successful design by October 1930, it offered the world's first diesel powered crawler in 1931.

EMERSON-BRANTINGHAM

Emerson-Brantingham Implement Co. of Rockford, Illinois, traces its origins to an early competitor of Cyrus McCormick, the Manny Reaper Co., founded in 1852. Ralph Emerson, a cousin of the famous poet, Ralph Waldo Emerson, bought into the firm and years later brought in Charles Brantingham. Brantingham became president of the firm in 1909. In 1912 E-B purchased a number of other companies, including the Gas Traction Co. of Minneapolis; Reeves & Co. of Columbus, Indiana; and Geiser Manufacturing Co. of Waynesboro, Pennsylvania. All three of these firms had begun making tractors. Production continued on Geiser's *20-hp Tractor* only for about a year. Reeves' four-cylinder *40-65* was built until 1920. Gas Traction's four-cylinder *Big 4 "30"* was produced until 1916. In 1913 E-B brought out two other Big 4 designs, the *Model D 20-35* and the *45-90*. The 45-90 was built until 1916 and the Model D until 1919.

An advertisement in the December 12, 1914 issue of *Progressive Farmer* announced the *Emerson Farm Tractor Model L 12-20*. This lightweight four-cylinder machine was of unique three-wheel design, featuring a single drum drive-wheel in the rear. Its production lasted through 1917. That year E-B introduced its *Model Q 12-20*. Also in 1917 an *E-B 9-16* model was announced. However, when the Fordson appeared in 1918, the 9-16 was evaluated and found to be inferior. It was dropped in 1920. In July 1918 E-B's *Model AA 12-20* made its debut to compete with Fordson. It weighed nearly a ton less than the Model Q but used the same engine. Production of both 12-20 models continued to 1928. The *E-B 20-35*, introduced in 1919, was an improved version of the Big 4 Model D. It was discontinued in 1920. E-B's *16-32* was introduced in 1921 and built to 1928. In 1923 the company brought out its *No. 101 Motor Cultivator*. It too remained in production until 1928. At one brief moment in its history E-B had been the world's largest farm implement company. But times changed, its tractors became obsolete, and in 1928 it was sold to the J.I. Case Co.

FORDSON

Henry Ford, founder of Ford Motor Co., had grown up on a farm, and beginning in 1907 had taken an interest in developing an inexpensive tractor. This process required ten long years. In the meantime W. Baer Ewing, a shady entrepreneur, decided to capitalize on

Ford's good reputation. In 1916 he employed a Paul B. Ford as designer, and established the Ford Tractor Co. Mainly intended as a stock swindle, the company was first established as a South Dakota corporation, but when it went bankrupt, a Delaware corporation was set up. A limited number of poor quality three-wheel tractors were built. However the company soon collapsed and Ewing fled to Canada.

This left Henry Ford in a situation where he had to choose another name for his tractor, and Fordson was selected. Fordson was the first to introduce mass-production techniques to the industry, and its first tractor came off the assembly line on October 8, 1917. Due to an urgent appeal from the British government, the first 7,000 units were diverted there. Thus it was only in early 1918 that the four-cylinder *Fordson Model F 9-18 Tractor* arrived on the American market. Backed by the Ford name, priced cheaply at $700, and having unit-frame construction, it was very appealing to farmers. Of a total American tractor production of 133,000 units that year, Fordson built 34,167, taking over the sales lead from International Harvester Co.

In subsequent years the farm economy collapsed, the tractor market became saturated, and competition sharply intensified. By 1922 the Model F sold for only $395. That year American tractor production totaled 98,724 units. Fordson's share was 62,000 units for 63% of the market. While Fordson dominated the market at this time it was often selling below cost to do so. By 1928 the Model F was obsolete and American manufacture ceased. Fordson production continued at a plant it had established in 1919 at Cork, Ireland. In 1929 it replaced the Model F with the *Model N*. All tractor production was moved to Dagenham, England, in 1933.

HART-PARR and OLIVER

In 1907 Hart-Parr dropped its earlier production designs and introduced the *30-60* model. Having oil cooling and two cylinders, it soon acquired the name of "Old Reliable." It was built until 1918, with production totaling over 4,100 units. From 1908 to 1914 Hart-Parr offered an even larger design, the two-cylinder *40-80*. Built from 1909 to 1912 was Hart-Parr's two-cylinder *15-30*. In 1911 and 1912 a few *60-100* tractors were built. These four-cylinder giants weighed over 25 tons.

Between 1912 and 1914 Hart-Parr offered its two-cylinder *20-40*. Featuring a transmission with two forward speeds, its total production reached 600 units. Appearing in

1914 was the 5 1/2 ton *12-27*. Modified slightly and re-rated in 1915, it became the *18-35 Oil King*. Over 2,000 units were built when production stopped in 1918. Also introduced in 1914 was Hart-Parr's first really lightweight model, the 2 1/2 ton two-cylinder *Little Devil 15-22*. Its unique design featured a single drive-wheel at the rear, with the driver positioned to the right. Having no differential and a two-cycle engine, it could be run at two speeds, both forward and backward. Unfortunately mechanical problems with this design caused grief for Hart-Parr. After 725 units had been built production halted in 1916. $500,000 was spent in a subsequent recall.

Hart-Parr also lost heavily after tooling up for a British Government munitions contract only to have it canceled. This resulted in a company reorganization in 1917. Charles Hart was removed as General Manager and other key people fired. Parr was moved to head of engineering. In February 1918 the firm introduced its *New Hart Parr 12-25*. Developed by Hart before his departure, this two-cylinder water-cooled tractor was shortly re-rated, becoming known as the *"Thirty"* or *15-30 Type A*. When production stopped in 1922 over 10,000 units had been built. However, with slight modifications this series continued until 1930. Subsequent versions were the *15-30 Type C* built to 1924, the *16-30 Type E* built to 1926, the *18-36 Type G* built to 1928, and the *18-36 Type H*. The Type E was the first Hart-Parr tractor to enclose the final drive. In 1927 the Type G was equipped with a three-speed transmission.

In 1921 and 1922 Hart-Parr offered its two-cylinder *"Twenty"* or *10-20 Type B*. A slightly changed *10-20 Type C* was built from 1922 to 1924. Priced at $920 in 1923, it offered little competition to its main rival, the Fordson. This series continued in the *12-24 Type E* built to 1928 and the subsequent *12-24 Type H*. The latter was equipped with a three-speed transmission. From 1923 to 1927 Hart-Parr built its *22-40*, powered by two combined 10-20 two-cylinder engines. A total of 496 units were built. Replacing it was the *28-50*. By 1929 the firm was in serious financial trouble. That year, along with Nichols & Shepard Co., American Seeding Machine Co., and Oliver Chilled Plow Works, it was merged into the Oliver Farm Equipment Co. Replacing the old lineup in 1930 were three four-cylinder models with unit-frame construction. These were the *Oliver-Hart-Parr 18-28*, its row-crop sibling the *18-27*, and the *Oliver-Hart Parr 28-44*. By mid-decade the Hart-Parr name was quietly dropped, while the Oliver name continued.

HEIDER and ROCK ISLAND

A farmer for thirty years, John Heider, along with two sons, established Heider Manufacturing Co. in Carroll, Iowa. In 1911 Heider brought out a lightweight tractor, the *10-20*, having a four-cylinder Waukesha engine. It used an infinitely variable friction drive system. Improvements on the tractor were made in 1912 and soon it was being marketed by the Rock Island Plow Co. This firm, based in Rock Island, Illinois, was one of the largest farm implement manufacturers in the country. Rock Island was very successful in selling the Heider and bought all rights to it in 1916.

That year a *Heider Model C 12-20* was introduced. Also having a four-cylinder Waukesha engine, it remained in production until 1924. It was replaced by an improved *Model C 15-27*. Also appearing in 1916 was the Waukesha powered *Heider Model D 9-16*.

Built between 1920 and 1926 was the *Heider 6-10 Cultivator*, powered by a four-cylinder LeRoi engine. When production of the Model D ended in 1929 the Heider name was dropped. Rock Island began building tractors under its own name in 1927, introducing the *Model F 18-35*. Powered by a four-cylinder Buda engine, it was built until 1937 when the firm was sold to the J.I. Case Co.

INTERNATIONAL HARVESTER

In 1902 two major farm implement manufacturers, McCormick Harvesting Machine Co. and Deering Harvester Co., plus three smaller implement firms were amalgamated into what instantly became the world's largest farm equipment producer, the International Harvester Co. Based in Chicago, the company built its first experimental tractor in 1905 and produced fourteen others the following year. In 1907 the company began standardized manufacture with its single cylinder *IHC Type A*. During its production run thru 1916 the Type A was offered in 15 and 20-hp variations. It had evaporative cooling, friction drive, and was successful for its time, selling almost 630 units in 1908. Similar to the Type A, the 20-hp *IHC Type B* was introduced in 1910 and produced thru 1917. These tractors were assembled at Upper Sandusky and Akron, Ohio.

Both McCormick and Deering had created extensive dealer networks prior to the merger, and were still maintaining them separately at this time. Thus after introduction of the IHC A and B types, it was decided to offer two separate tractor lines; the *Mogul* for exclusive sale by McCormick dealers and the *Titan* for Deering. In 1909 the *20-hp Mogul Type C* was introduced and in 1910 the *20-hp Titan Type D*. The Mogul Type C upgraded to 25-hp in 1911 and the Titan Type D in 1912. Production for both models ceased in 1914. By 1911 IHC had become the primary manufacturer of farm tractors. It maintained this position until challenged by Fordson in 1918.

International Harvester also began building tractors in Milwaukee and Chicago. Titan models were mainly produced in Milwaukee while Moguls were built in Chicago. Among the first from the Chicago plant was the two-cylinder *Mogul 45*, introduced in 1911. Revised in 1912, it became the *Mogul 30-60*. When production ceased in 1917, approximately 2,437 of these machines had been built. International Harvester was constantly improving the technology and a number of other Mogul and Titan models were manufactured.

In 1914 IHC began competition in the lightweight tractor market, introducing its *Mogul 8-16*. Selling 14,065 units by 1917, this single-cylinder tractor was replaced in favor of the new *Mogul 10-20*. Similar in appearance to its predecessor, the single-cylinder 10-20 was built from 1916 through 1919. Over 8,900 machines were produced. The two-cylinder *Titan 10-20* was built from 1916 to 1922. It was extremely popular and over 78,000 units were sold.

The *International 8-16*, originally intended as a Mogul model, made its debut in 1917. More modern in appearance than other IHC tractors, its four-cylinder engine was very similar to that of International's Model G Truck. Adopting a design feature of the truck, it had a sloping hood and was cooled by a radiator positioned behind the motor. Three successive versions were made, and in 1919 the 8-16 became the first American tractor to offer an optional *power take-off* or PTO. This allowed a transfer of tractor power to run a machine being pulled behind. When 8-16 production ceased in 1922, approximately 33,138 units had been built.

Surviving against the Fordson required International Harvester to become seriously competitive. It cut prices and began to offer free plows with purchase of its Titan 10-20. However by 1922 IHC market share had shrunk to 12%, with the company building only 11,781 tractors. Production of the Titan 10-20 and International 8-16 was discontinued. Beginning the previous year IHC started to counter Fordson by offering a vastly improved model, the *McCormick-Deering 15-30*. Weighing three tons, it was the first IHC model to have unit-frame construction. Numerous options were offered including rear power takeoff. In 1929 this tractor was upgraded, becoming the *Model 22-36*. Its production ended in 1934.

In 1922 the *McCormick-Deering 10-20* made its debut. It was very similar to the 15-30 but weighed one ton less. A very popular model, it was manufactured to 1939, with production totaling over 215,000 units. IHC market share increased from 9% in 1923 to 16% in 1924, reaching 29% in 1926, and 55% in 1928.

Years of careful development went into the revolutionary *McCormick-Deering Farmall*. Not only intended as the ultimate row-crop machine, it could also be used for normal plowing and belt work. Built high off the ground to clear crops, it featured narrow front wheels for running between rows. It was designed for use with many types of implements, which could be mounted in the front or rear. However IHC was cautious in introducing it, because a publicity campaign for the 10-20 was underway. Thus the amazingly versatile Farmall was quietly introduced in 1924, and word slowly got out. Only 200 were manufactured that year, but annual production increased to 4,430 in 1926, 9,502 in 1927, and 24,899 in 1928. Despite the onset of the Great Depression a crescendo was reached in 1930 when 42,093 units were built.

JOHN DEERE

In 1912, years after giving up on John Froelich's early tractor, Waterloo Gasoline Engine Co. of Waterloo, Iowa, took another chance. That year it introduced its *Waterloo Boy One-Man Tractor* designed by Harry Leavitt. Featuring a four-cylinder cross-mounted engine, its production ended in 1913. A *One-Man Tractor* of different design followed and was built until 1914. It featured a 15-hp two-cylinder engine.

The two-cylinder *Waterloo Boy Style R* was built from 1914 to 1918. It had a sideways-mounted cellular radiator. Originally rated at 12-24, it went through a number of modifications and was later re-rated at 12-25. The popular two-cylinder *Style N* was manufactured from 1917 to 1924. Rated at 12-25, it was equipped with a two-speed transmission. Nearly 20,000 units were built.

In 1918 Waterloo was sold to Deere & Co. of Moline, Illinois. An old and respected implement firm, Deere had been conducting tractor experiments since 1912. However, none of its designs had ever gone into production. In purchasing Waterloo it was suddenly in a very good position. While production of the Style N continued, Deere took another Waterloo design and perfected it. This became the famous *John Deere Model D Tractor*, introduced in 1924. Although differing significantly in appearance, it retained much of its predecessor's solid engineering. It featured a side-by-side horizontal two-cylinder engine, rated at 15-27. Becoming famous for its ruggedness and simplicity, the Model D went through various modifications with production lasting to 1953. Around 160,000 of these tractors were built. John Deere's *Model GP* was introduced in 1928. In late 1929 the company first advertised the *GP Wide-Tread Tractor*, its first row-crop design. It was built to 1935.

RUMELY

The M. Rumely Co. of La Porte, Indiana, noted for its threshers and steam traction engines, traced its roots back to 1852. Dr. Edward A. Rumely, grandson of the company's founder, studied medicine in Germany, and while there became acquainted with Rudolf Diesel. Influenced by Diesel's ideas he returned determined to develop a tractor for his family's business. For this project he recruited John A. Secor, who since 1885 had been experimenting with low-grade fuel engines, and Rumely's shop superintendent William H. Higgins. Experimental prototypes built in 1908-09 were so successful that the company moved quickly to begin tractor production. Rumely's Oil Pull tractors had a distinctive design and were cooled by oil, using an induced draft system.

The first production model, the two-cylinder Type B 25-45, was produced from 1910 to 1914. It was quickly joined by the two-cylinder Type E 30-60, built from 1910 to 1923, and the one-cylinder Type F 15-30, manufactured from 1911 to 1917. These big tractors quickly gained a reputation for their excellence and reliability. During its period of manufacture the Type E sold 8,224 units. Good profits at this time allowed Rumely to buy out two competitors, Gaar, Scott & Co. of Richmond, Indiana, and Advance Thresher Co. of Battle Creek, Michigan in 1911. The following year it bought out Northwest Thresher Co. of Stillwater, Minnesota. With the latter two acquisitions, Rumely gained a new tractor model. It was the Universal Farm Tractor built both by the Universal Tractor Co. of Stillwater, Minnesota, and American-Abell Engine & Thresher Co. of Toronto. It was subsequently marketed as the Rumely Gas Pull 20-40. Later down rated to 15-30, it was sold until 1915. Rumely also briefly offered a Gas Pull 35-70 model in 1913.

Bad luck befell Rumely after it sold too many tractors in Canada for the 1914 season. A crop failure that year left the company in serious financial trouble. Going into receivership in early 1915, it was reorganized as the Advance-Rumely Thresher Co., with the Rumely family no longer in control. Rumely's first venture into the small tractor market came in 1916 when it offered the *All-Purpose Tractor*. Having one forward drive wheel, a forward outrigger-style support wheel, and a rear wheel for steering, it came in 8-16 and 12-24 versions. This awkwardly styled model was only manufactured for a couple of years.

In order to stay competitive Rumely continued to offer new models, bringing out its "Heavyweight" line in 1918 and its "Lightweight" line in 1924. The 1918 series included the *Type G 20-40*, *Type H 16-30*, and *Type K 12-20*. The 1924 series was of more compact style and offered a new engine design. It included the *Type L 15-25*, *Type M 20-35*, *Type*

R 25-45, and *Type S 30-60*. In 1924 Rumely bought out a major competitor, Aultman & Taylor Machinery Co. of Mansfield, Ohio.

In 1927 Rumely purchased the farm tractor line of Toro Motor Co. of Minneapolis. This company's motor cultivator was modified to become the *Rumely 20-hp DoAll Tractor* introduced in 1928. Also brought out in 1928 and 1929 was an updated version of the "Lightweight" series of 1924. This group included the *Type W 20-30*, *Type X 25-40*, *Type Y 30-50*, and *Type Z 40-60*. Showing a new design direction was the *Model 6A 28-43* of 1930. However, foreign sales losses and the onset of the Great Depression left Rumely in a precarious state. It was sold to Allis-Chalmers in 1931.

SAMSON

In 1917 General Motors Corporation, under William C. Durant, was engaged in fierce competition with Ford Motor Co. With the Fordson Tractor about to enter the market GMC decided to enter the tractor business as well. To reduce product development time, it opted to buy out an already existing firm, the Samson Sieve Grip Tractor Co. of Stockton, California. This company had begun as the Samson Iron Works in 1884 and became involved with internal combustion engines around 1904.

About 1914 it entered the tractor market, and from then through 1917, produced two models of its long, low-slung, three wheel design. These were the single-cylinder *4-5 Horsepull 10 bhp Sieve Grip*, later designated as the *6-12*, and the four-cylinder *8-10 Horsepull 20 bhp Sieve Grip*, later designated as the *10-25*. An early advertisement for another Samson model, the four-cylinder *12-25 Sieve Grip*, appeared in the March 24, 1917 issue of California Cultivator. The term "Sieve Grip" referred to the tread of the open-faced rear wheels, which apparently gave better traction on certain kinds of soil.

In 1916 the company changed its name to Samson Tractor Co. In 1917 the name was changed again to Samson Sieve Grip Tractor Co. After the GMC purchase much of Samson's operation was moved in 1918 to a facility in Janesville, Wisconsin. There the newest model was given a four-cylinder GMC engine and other improvements made. Production of the *GMC Samson Sieve Grip 12-25* apparently ran into 1919, the same year that the Stockton facility was closed. However, it offered no competition for the Fordson and so a new design was announced in December 1918.

Shorter, lighter, and more conventional in style, the four-cylinder Samson Model M made its appearance the following year. Rated at 12-20, it weighed 3,300 pounds, had a 276 cubic inch Northway engine, and was competitively priced at $650. In comparison the Fordson was rated at 9-18 and weighed 2,700 pounds. Governor, fenders, and belt pulley were extras on the Fordson, but standard on the Samson.

A larger *Samson Model A* was announced in early 1919 but never built. About the same time the *Samson Model D Iron Horse* was announced. Having acquired the rights to the Jim Dandy Motor Cultivator, GMC gave it a four-cylinder Chevrolet engine and made other modifications, offering it for sale at $450. Unfortunately the Model D was poorly designed and had little appeal to farmers. In the tight tractor market that followed, Fordson responded to the Model M and other competition by drastically reducing prices. Losing heavily by 1922, GMC was forced by its stockholders to leave the tractor business.

TWIN CITY and MINNEAPOLIS-MOLINE

Minneapolis Steel & Machinery Co. of Minneapolis, Minnesota was formed in 1902. In 1910 the company hired the Joy Wilson Co. of Minneapolis to develop a large four-cylinder tractor. Five prototypes were built. Pleased with the result, MS&M contracted out to the J.I. Case Threshing Machine Co. production of a redesigned version, the *Twin City 40*. Five hundred units were built in 1911 and 1912. An improved version of the 40 was produced from 1913 to 1924. MS&M began building tractor motors for Reeves & Co. It also contracted in late 1913 to build tractors for Bull Tractor Co. Other Twin City models were offered in 1913. These were the *15*, built to 1917, and the *25* and *60*, built to 1920. The 60, rated at 60-90, weighed 14 tons, had a six-cylinder engine, and was water-cooled. The other models had four-cylinder motors. Three different versions of the 15 were built.

In 1918 a radically different lightweight design was introduced. This was the 16-30, which had a totally enclosed chassis and a long, low-slung appearance. It featured a four-cylinder L-head engine and two-speed transmission. Production ran to 1920. More conventional designs followed, with an important engineering innovation, a four-cylinder overhead-valve engine with four valves per cylinder. It came standard in the *12-20* built from 1919 to 1927, and the *20-35* produced between 1920 and 1929. The 20-35 was essentially a larger version of the 12-20.

Twin City's *Type FT 21-32* was produced from 1926 to 1934. The *17-28 Model TY*, built from 1927 to 1934, was a re-rated version of the 12-20. The *27-44 Model AT*, produced between 1929 and 1935, was a re-rated version of the 20-35. In April 1929 MS&M was merged along with another tractor builder, Minneapolis Threshing Machine Co., and the Moline Plow Co., into the Minneapolis-Moline Power Implement Co. of Minneapolis. Under Minneapolis-Moline the Twin City brand continued for some years, with a number of new models offered.

WALLIS and MASSEY-HARRIS

The first *Wallis Bear* tractor was built by Wallis Tractor Co. of Cleveland in 1902. Although remarkably engineered, only perhaps nine 60-hp Bear models were built up to 1910. Other designs followed, among them the *Fuel Save* model. H.M. Wallis, president of the firm was the son-in-law of J.I. Case, and later became president of J.I. Case Plow Works of Racine, Wisconsin, an entirely separate company from J.I. Case Threshing Machine Co., of the same city. By 1913 Wallis had moved his own operation to Racine, and setting it up at the Plow Works, introduced his revolutionary *Cub* model. Having a modern lightweight three-wheel design, it was the first tractor ever to be built with unit-frame construction. This model was followed in 1915 with the four-cylinder *Model J 13-25* or *Cub Junior*. In this tractor the final drive gears were enclosed as well, fully protected from dust and dirt. This innovation was soon copied by Henry Ford for the design of his Fordson Model F tractor.

While the Wallis operation had been run separately up to this time, it was incorporated into the Plow Works in 1919. That same year the four-wheel *Model K 15-25* appeared and was built until 1922. Also introduced in 1919 was the *12-hp Motor Cultivator*. However, it was only built for a short time. With the intensified competition of the early 1920's, Wallis was able to stay in business due to its reputation for quality engineering. In 1922 it introduced its *Model OK 15-27*. It was replaced by the *20-30* in 1927. Orchard versions of these two models were also sold. In 1927 the Massey-Harris firm of Toronto began selling Wallis tractors in Canada. The following year it bought out the J.I. Case Plow Works, but keeping the tractor operation, sold back rights of the Case name to J.I. Case Threshing Machine Co.

Massey-Harris had twice failed in efforts to enter the tractor business. It first did so in 1917 by marketing the Big Bull tractor in Canada. In the early 1920s it sold three models built by Parrett. In acquiring Wallis it finally gained solid traction. Massey-Harris incorporated in Racine and built a plant there. Production of the standard and orchard 20-30 Wallis models continued into 1931. The *Wallis 12-20*, introduced in 1929, was built to about 1933. The *Massey-Harris General Purpose 15-22* was introduced in 1930. Subsequent Massey-Harris designs incorporated the boilerplate unit-frame patented by Wallis. In 1953 Massey-Harris merged with Harry Ferguson Inc. to become Massey-Ferguson Inc.

British Tractors During and After World War I

At the onset of World War I Britain's farm economy was depressed because cheap agricultural products were imported from her colonial empire. At this time there were about fifteen domestic tractor manufacturers. Britain's circumstances suddenly worsened when men and draft animals were sent off to war and German U-boats threatened its overseas food supply. As factories turned to war production, tractor manufacture ceased. The only major company continuing was Saunderson. Serious government efforts began in 1916 to increase farm productivity. Women and disabled veterans were recruited for farm employment, fallow land was plowed, and a sizeable number of American tractors were imported.

While IHC's Mogul 8-16 and Titan 10-20 were sold through local company branches, British firms marketed other tractors. America's Parrett tractor was sold as the *Clydesdale*, the Big Bull became the *Whiting Bull*, and the Peoria was marketed as the *Culti-Tractor*. The Waterloo Boy became the *Overtime*. Upon arrival in Britain, Waterloo tractors were repainted and given new markings and serial numbers. About 4,000 were sold. In late 1917 seven thousand Fordson tractors arrived from the U.S.

After the war factories were converted to peacetime use, but much had changed. New tractor manufacturers emerged. Among these was Peter Brotherhood Ltd. of Peterborough. From 1919 to 1930 it built the *Peterbro 18-35*. Another firm was D.L. Motor Manufacturing Co. of Motherwell. From 1919 to 1924 it built the three-wheeled *Glasgow Tractor*. All three wheels received power from a 27-hp Waukesha engine. Only about 400 were built, however. For awhile Moseley Motor Works of Birmingham built the *Garner 30-hp Tractor*. Perhaps the smallest of the postwar machines were the *Crawley Motor Plow* and the *Weeks-Dungey New Simplex Model*. The Weeks-Dungey was designed especially for hop cultivation. The *Blackstone Tractor*, built at Stamford, was offered in wheeled and crawler versions. The company's three-cylinder engine became known for its smoothness. Another crawler producer was Clayton & Shuttleworth of Lincoln. Its *Clayton-Shuttleworth 18-27* was built from 1916 to 1928.

The most successful postwar tractor firm was Austin Motor Co. From 1919 to 1926 it built its *Austin 25-hp Tractor* at Longbridge, Birmingham. Three thousand units had been built by August 1922, but Fordson competition eventually forced Austin out of Britain. However, production continued in France. Under license, Ruston of Lincoln built the *British Wallis* from 1920 to 1929. In a similar arrangement, Vickers of Newcastle-upon-Tyne manufactured the McCormick-Deering 15-30. Built from 1925 to 1933, it was called the *Aussie*, and as suggested by its name, many were exported to Australia. Late in the decade the firms Garrett of Leiston, and J&H McLaren of Leeds, built a few diesel-powered tractors.

Unfortunately, in the depressed economy of this period British tractors didn't sell well, and production durations were often brief. The wartime importation of tractors established an American presence that was to last. Some very good ideas were explored in domestic tractor design, but British companies couldn't compete with the large-scale operations of American firms, especially Fordson. Fordson's position was strengthened in 1919, when mass-production of the Model F began at Cork, Ireland. Inspired by the engineering genius of Harry Ferguson, Britain's tractor industry began to revive somewhat in the 1930s.

Traction Engines,
Cable Plowing
and
Early Farm Tractors

Traction Engines:

Top: Boydell's "Steam Horse" traction engine, Messrs. Charles Burrell, Thetford, England, 1857.

Middle: Two-cylinder Boydell traction engine shipped to Brazil, 1860.

Bottom: J.W. Fawkes steam traction engine demonstrated in Philadelphia, July 20, 1859.

Traction Engines & Cable Plowing:

Top: Two Engine System using equipment of the John Fowler Co. of Leeds, c.1870.

Middle Left: Cable winding engine with windlass for Roundabout System, Barford & Perkins, Peterborough, England, 1870s.

Middle Center: Anchor cart, c.1870.

Bottom: Cable winding engine of Messrs. Yarrow & Hilditch, 1862

Traction Engines & Cable Plowing:

Top: One Engine System, c.1870.

Bottom: Cable winding engine of the Lotz Co., Nantes, France, C.1870s.
It had two vertical winding drums that girdled the boiler.

E. BOURDELIN.

CHARLOT

Traction Engines:

Top: Steam traction engines of Aveling & Porter Co., Rochester, England, C.1870s.

Middle Left: Unidentified American traction engine, c.1885.

Bottom: Steam traction engine, Gaar Scott & Co., Richmond, IN, 1882.

Traction Engines:

Top: Ingleton's Automatic Track Engine, Ingleton Steam Plow Co., Pottstown, PA, 1895.

Bottom: Advance Traction Engine, Case & Willard Thresher Co., Battle Creek, MI, 1880s.

Traction Engines:

Top: Peerless Traction Engine pulling plow, Geiser Manufacturing Co., Waynesboro, PA, 1903.

Middle: Traction engine powering threshing machine, c.1900.

Bottom: Side crank, spring mounted traction engine, J.I. Case Threshing Machine Co., Racine, WI, 1899.

Traction Engine/Tractor:

A small number of experimental kerosene powered traction engines were built in the late nineteenth century. This machine, powered by an Akroyd-Stuart engine, was built by Richard Hornsby & Co., Grantham, England, 1896.

Early Tractors:

Hart-Parr Tractors, Hart-Parr Co., Charles City, IA.

Top: 1907-18 Model 30-60 "Old Reliable."

Bottom: 1912-14 Model 20-40. In the system devised by the tractor
industry for indicating power, the first number indicated drawbar horsepower
and the second indicated belt-pulley horsepower.

Early Tractors:

Top: 1910-23 Rumely Type E 30-60 Oil Pull Tractor, M. Rumely Co., La Porte IN.

Bottom: 1910-14 Fairbanks-Morse 15-25 Tractor, Fairbanks, Morse & Co., Chicago, IL.

Early Tractors:

Top Left: 1911-13 Quincy Model "O" 20-30 Tractor, Electric Wheel Co., Quincy, IL.

Top Right: 1913-16 IHC Titan 18-35 Tractor, International Harvester Co., Chicago, IL.

Middle Left: 1913-15 Kuhnert's Vanadiumized Tractor, L.C. Kuhnert & Co., Chicago, IL.

Middle Right: 1914-17 IHC Titan 12-25 Tractor.

Bottom: 1910-14 IHC Titan Type D 20 h.p. Tractor.

KUHNERT'S
VANADIUMIZED
TRACTOR

TITAN

Early Tractors:

Top: 1912-17 IHC Mogul 30-60 Tractor, International Harvester Co., Chicago, IL.

Middle left: 1912 Flour City Tractor, Kinnard Haines Co., Minneapolis, MN.

Bottom: 1912-18 IHC Mogul 12-25 Tractor.

Early Tractors:

Top: 1912-16 Big Four "30," Tractor, Emerson-Brantingham Co., Rockford, IL.

Middle left : 1911-12 Heider 10-20 Tractor, Heider Mfg. Co., Carroll, IA.

Bottom: 1913-24 Twin City "40" 40-65 Tractor, Minneapolis Steel & Machinery Co. Minneapolis, MN.

Early Tractors:

Top: 1913-15 Leader 12-18 Tractor, Leader Engine Co., Detroit, MI.

Upper Middle Left: 1912-16 Heer four-wheel drive Tractor, Heer Engine Co., Portsmouth, OH.

Middle Center: 1913-20 Nichols-Shepard 35-70 Tractor, Nichols & Shepard Co., Battle Creek, MI.

Middle Right: 1915-17 Emerson Model L 12-20, Emerson-Brantingham Co., Rockford, IL.

Lower Middle Left: 1914-15 Wadsworth 18 h.p. Tractor, Detroit Engine Works, Detroit, MI.

Bottom: 1911-14 Hackney Auto Plow, Hackney Mfg. Co., St. Paul, MN.

Early Tractors:

Top Left: 1914-15 Hart-Parr 12-27 Tractor, Hart-Parr Co., Charles City, IA.

Top Right: 1911-(?) Flour City 30-50 Tractor, Kinnard Haynes Co., Minneapolis, MN.

Middle Left: 1914-16 Peoria 8-20 Tractor, Peoria Tractor Co., Peoria, IL.

Middle Right: 1913-14 Waterloo Boy 16 h.p. two-cylinder "One-Man" Tractor,
Waterloo Gasoline Engine Co., Waterloo, IA.

Bottom: 1914-27 Russell Giant 40-80 Tractor, Russell & Co., Massilon, OH.
In 1921 it was down rated to 30-60.

Early Tractors:

Crawler tractors:

Top: c.1909 Holt Caterpillar Model 40, Holt Mfg. Co., Stockton, CA.

Middle Left: 1913-(?) Best "75," C.L. Best Gas Traction Co., San Leandro, CA.

Middle Right: 1913-17 Strait's 30-50 Tractor, Killen-Strait Mfg. Co., Appleton, WI.

Bottom: 1915 Bullock "Baby" Creeping Grip Tractor, Bullock Tractor Co., Chicago, IL.

Early Tractors:

Top: 1912-15 Rumely 20-40 Gas Pull Tractor, M. Rumely Co., La Porte IN. The rights to this tractor previously known as the Universal Farm Tractor, were acquired when Rumely purchased both Advance Thresher Co. in 1911 and Northwest Thresher Co. in 1912. This design was jointly manufactured by Universal Tractor Co. of Stillwater, Minnesota, a Northwest subsidiary, and American-Abell Engine & Thresher Co. of Toronto, co-owned by Advance and Minneapolis Threshing Machine Co. Minneapolis Threshing Machine Co. received a payment of cash for its share of the rights.

Bottom: 1914-20 Sandusky Model E 15-35 Tractor, Dauch Mfg. Co., Sandusky OH.

Early Tractors:

Top Left: 1913-15 Wallis Cub Tractor, J.I. Case Plow Works, Racine WI.
This was the first tractor with unit frame construction.

Top Right: 1914-16 Hart-Parr "Little Devil" 15-22 Tractor, Hart-Parr Co., Madison, WI.

Upper Middle Right: 1914-18 Case 12-25 Tractor, J.I. Case Threshing Machine Co., Racine, WI.

Middle Left: 1912-16 Lambert Steel Hoof Tractor, Lambert Gas Engine Co., Anderson IN.

Lower Middle Right: 1915-18 Wallis Cub Jr. Model J 13-25 Tractor.

Bottom: 1916?-17 Twin City "15" 15-30 Tractor, Minneapolis Steel & Machinery Co.,
Minneapolis, MN. This was the final of three versions of the "15" model that began production in 1913.

Early Tractors:

Top: 1914-17 one-cylinder Samson 4-5 Horsepull 10 bhp Sieve Grip Tractor, Samson Iron Works, Stockton, CA. This model was later designated as the 6-12.

Middle: 1914-17 four-cylinder Samson 8-10 Horsepull 20 bhp Sieve Grip Tractor. This model was later designated as the 10-25.

Bottom: 1914-17 IHC Mogul 8-16 Tractor, International Harvester Co., Chicago.

Early Tractors:

Top four: Tractors of Avery Co., Peoria, IL:

Top Left: 1912-19 Model 12-25.

Top Right: 1913-20 Model 40-80.

Middle Left: 1918 5-10 Model B.

Middle Right: 1914-22 Model 8-16.

Bottom: 1916-22 Eagle Model F Tractor, Eagle Mfg. Co., Appleton, WI.
At this time the Model F was made in two versions, the 12-22 and the 16-30.

Early Tractors:

Top: 1916-18 Case 9-18 Tractor, J.I. Case Threshing Machine Co., Racine, WI.

Middle Left: 1916-19 Galloway's Farmobile 12-20, William Galloway Co., Waterloo, IA.

Middle Right: 1915-17 Big Bull 7-20 Tractor, Bull Tractor Co., Minneapolis, MN.

Bottom: 1915-20 Case 10-20 Tractor.

Early Tractors:

Top: 1915-17 Moline Universal Tractor, Moline Plow Co., Moline IL.

Middle: Final version of the Samson Sieve Grip 10-25 Tractor, Samson Tractor Co., Stockton CA. This image came from an ad in the February 3, 1917 issue of *California Cultivator*, which ran shortly before Samson's new 12-25 Sieve Grip model was first advertised.

Bottom: 1915-17 Rumely 8-16 Tractor, Advance-Rumely Thresher Co., La Porte IN.

Early Tractors:

Top: During World War I, British tractor production virtually halted due to the pressing need for war materials. American tractors, such as this Waterloo Boy Style R, filled the void. Imported by L.J. Martin, it became the Overtime Tractor. Upon arrival in Britain these were repainted and given new markings and serial numbers. This image is from an Overtime ad in Britain's *Mark Lane Express*, Sep. 4, 1916.

Bottom: 1916-18 Rumely Type F 18-35 Oil Pull Tractor, Advance-Rumely Thresher Co., La Porte, IN.

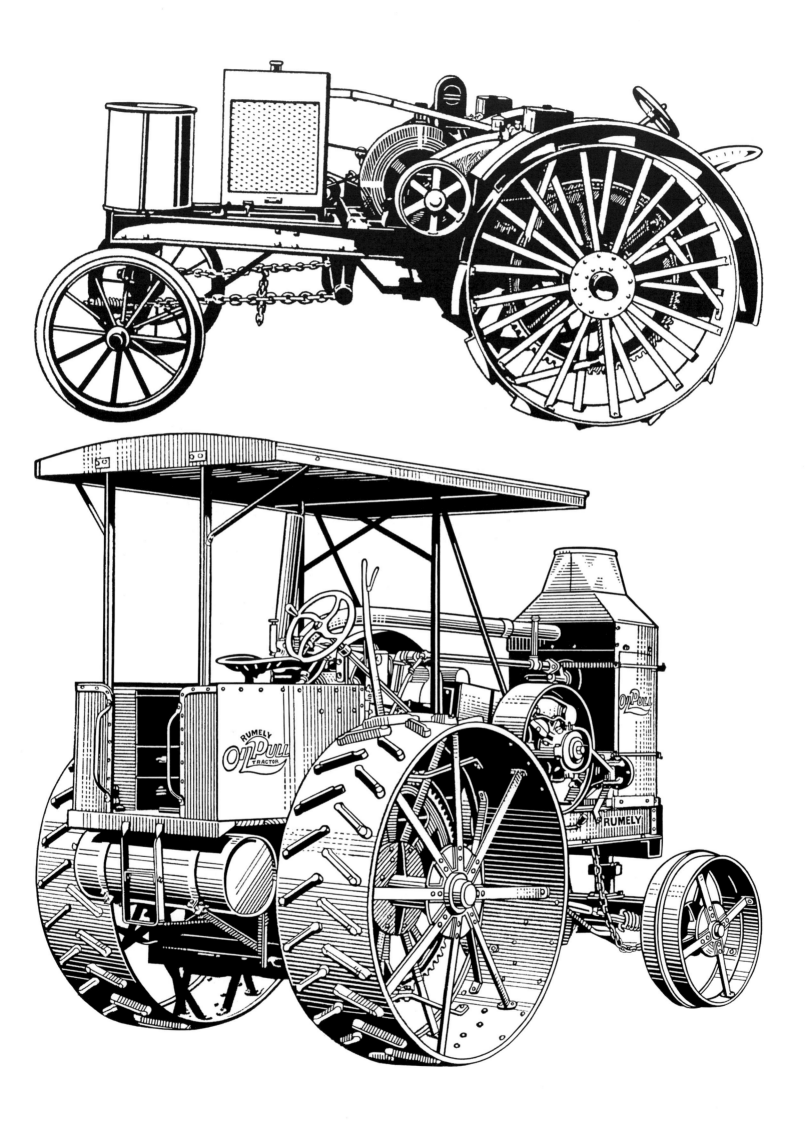

Early Tractors:

Crawler Tractors:

Top: 1917 Lambert Chain Tread Tractor, Lambert Gas Engine Co., Anderson, IN.

Bottom three: Yuba Ball Tread Tractors, Yuba Mfg. Co., Maryville, CA.

Middle Left: 1916-21 Model 12-20.

Middle Right: 1916-21 Model 20-35.

Bottom: 1914-15 Model 18-36.

Early Tractors:

. Top three: Different views of 1916-22 IHC Titan 10-20 Tractor,
International Harvester Co., Chicago, IL.

Bottom: 1916-19 Case 20-40 (late style) Tractor, J.I. Case Threshing Machine Co., Racine, WI.

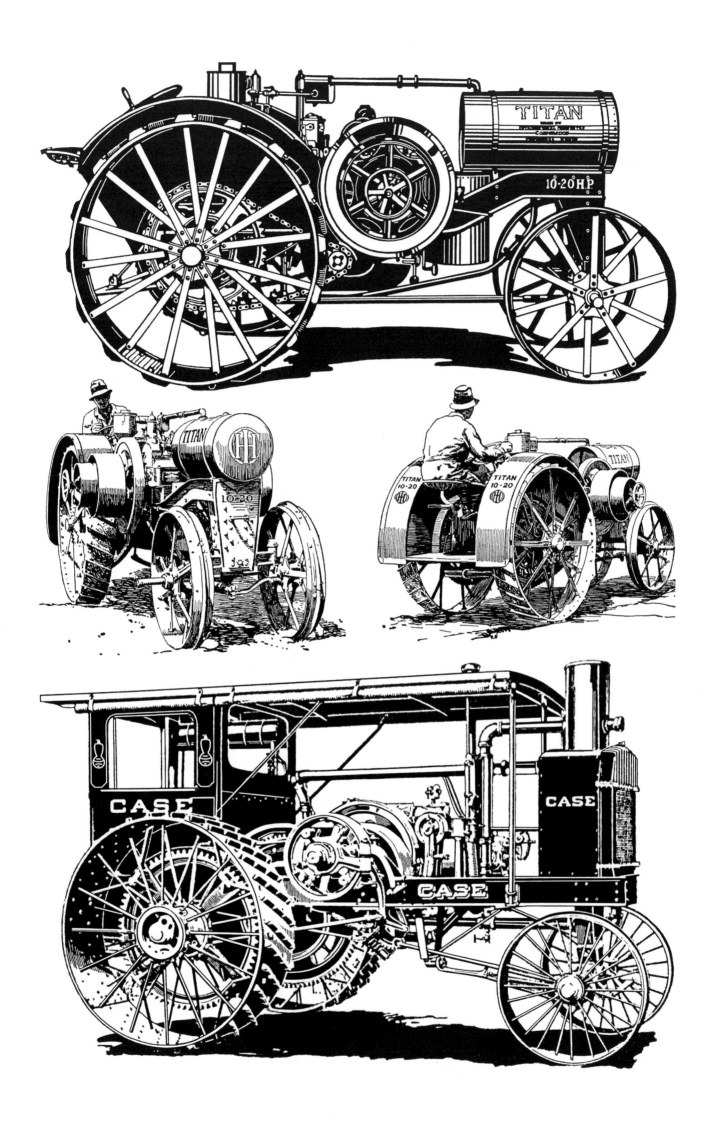

Early Tractors:

Top Left: 1918-21 Dart Blue J Tractor, Dart Truck & Tractor Corp., Waterloo, IA.

Top Right: 1916- 29 Heider Model D 9-16 Tractor, Rock Island Plow Co., Rock Island, IL.

Middle Right: 1918-22 Port Huron 12-25 Tractor,
Port Huron Engine & Thresher Co., Port Huron, MI.

Bottom: 1916-24 Heider Model C 12-20 Tractor

Early Tractors:

Top: 1917-28 E-B Model Q 12-20 Tractor, Emerson-Brantingham Co., Chicago, IL.

Bottom: 1917-20 Sandusky Model J 10-20 Tractor, Dauch Mfg. Co., Sandusky, OH.

Early Tractors:

Top: 1918-20 La Crosse Model G 12-21 Tractor, La Crosse Tractor Co., La Crosse, WI.

Bottom: 1916-29 Huber Light Four 12-25 Tractor, Huber Mfg. Co., Marion, OH.

Early Tractors:

Top: 1917-24 Appleton 14-28 Tractor, Appleton Mfg. Co., Batavia, IL.

Middle and Bottom: Two views of 1917-22 International 8-16 Tractor, International Harvester Co., Chicago, IL.

Early Tractors:

Top: 1917-18 Lauson 15-25 Tractor, John Lauson Mfg. Co., New Holstein, WI.

Upper Middle Center: 1917-20 Prairie Dog Model L 9-18 Tractor,
Kansas City Hay Press Co., Kansas City, MO.

Middle Right: 1918-20 Parrett Model H 12-25 Tractor, Parrett Tractor Co., Chicago, IL.

Bottom: 1919-25 Lauson 15-35 Tractor.

Early Tractors:

Top: 1918-21 Beaver 12-24 Tractor, Goold, Shapley & Muir Co., Brantford, Ontario.

Middle Left: 1918-(?) Uncle Sam 20-30 Tractor, U.S. Tractor & Machinery Co., Menasha, WI.

Middle Right: 1917-28 E-B Model Q 12-20 Tractor, Emerson-Brantingham Co., Chicago, IL.

Bottom: 1918-20 Turner Simplicity 14-25 Tractor, Turner Mfg. Co., Port Washington, WI.

Early Tractors:

Top Left: 1917-24 Rumely Oil Pull Type H 14-28 Tractor. This was later rerated to 16-30, Advance-Rumely Thresher Co., La Porte, IN.

Top Right: 1918-24 Rumely Oil Pull Type K 12-20 Tractor.

Middle Right: 1914-17 Simplex 15-30 Tractor, Simplex Tractor Co., Minneapolis, MN.

Bottom: 1919-21 Illinois 18-30 "Super Drive" Tractor, Illinois Silo & Tractor Co., Bloomington, IL.

Early Tractors:

Top: 1918-24 Fageol 9-12 Tractor, Fageol Motors Co., Oakland, CA.

Middle Left: 1917-19 Atlas 16-26 Tractor, Lyons-Atlas Co., Indianapolis, ID.

Middle Center, Middle Right, Bottom: Four-wheel drive tractors:

Middle Center: 1918-30 Fitch Four Drive 20-35, Four Drive Tractor Co., Big Rapids, MI.

Middle Right: 1918 Topp-Stewart 30-45, Topp-Stewart Tractor Co., Clintonville, WI.

Bottom: 1919 Nelson Tractor, Nelson Blower & Furnace Co., Boston, MA.

Early Tractors:

Top: 1918-28 E-B Model AA 12-20 Tractor, Emerson-Brantingham Co., Chicago, IL.

Middle Left: 1916-23 Allwork 14-28 Tractor, Electric Wheel Co., Quincy, IL.

Middle Right: 1917-20 Wichita 8-16 Tractor, Wichita Tractor Co., Wichita, KS.

Bottom: 1919-22 Wallis Model K 15-25 Tractor, J.I. Case Plow Works, Racine, WI.

Early Tractors:

Top: 1919-20 National Tractor, National Tractor Co., Cedar Rapids, IA. It was offered in 9-16 and 12-22 models. National was a subsidiary of General Ordnance Co., and later became the G-O Tractor.

Middle: 1917-19 Hession 12-24 Tractor, Hession Tractor & Tiller Co., Buffalo, NY.

Bottom: 1919-20 Coleman 16-30 Tractor, Coleman Tractor Corp., Kansas City, MO.

Early Tractors:

Top: 1920 Thorobred 18-30 Tractor, Commonwealth Tractor Co., Chicago, IL.

Middle Right: 1916-19 Plow Man 15-30 Tractor, Interstate Engine & Tractor Co., Waterloo, IA.

Bottom: 1918-24 Aultman-Taylor 15-30 Tractor, Aultman & Taylor Machinery Co., Mansfield, OH.

Early Tractors:

Top Left: 1917-24 Waterloo Boy Style N Tractor, Waterloo Gasoline Engine Co., Waterloo, IA; from 1918 on, Deere & Co., Moline, IL.

Top Right: 1918-22 Hart-Parr "30," or Type A 15-30 Tractor, Hart-Parr Co., Charles City, IA.

Middle Left: 1921 Townsend 25-30 Tractor, Townsend Mfg. Co., Janesville, WI. This tractor was designed to resemble a steam traction engine.

Middle Right: 1919-22 Samson Model M 12-20 Tractor, General Motors Corp., Detroit, MI.

Bottom: 1916-25 Russell Jr. 12-24 Tractor, Russell & Co., Massilon, OH.

Early Tractors:

Top: 1917-24 Waterloo Boy Style N Tractor. Waterloo Gasoline Engine Co., Waterloo, IA;
from 1918 onwards, Deere & Co., Moline, IL.

Bottom: 1919-24 Case 15-27 Tractor, J.I. Case Threshing Machine Co., Racine, WI.

Early Tractors:

Two views of 1918-22 Case 10-18 Tractor, J.I. Case Threshing Machine Co., Racine, WI.

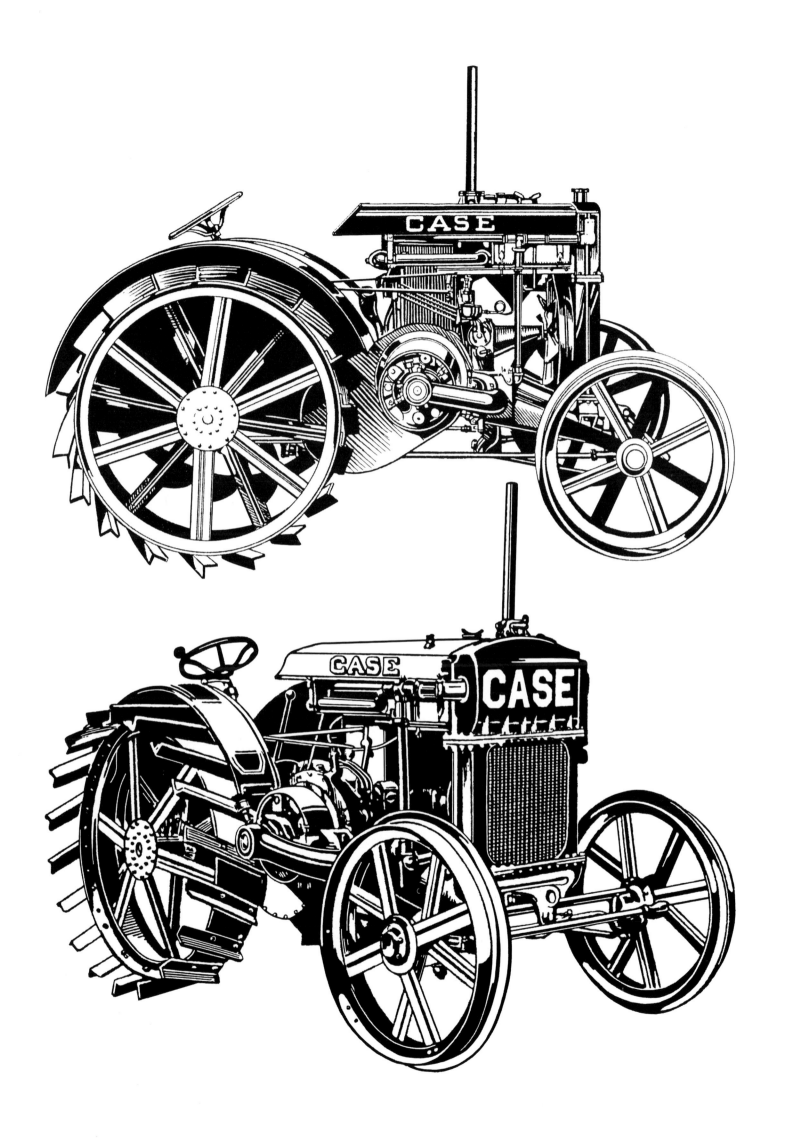

Early Tractors:

Top: 1919-23 Allis-Chalmers 6-12 Tractor, Allis-Chalmers Mfg. Co., Milwaukee, WI.

Middle: 1919 Moline Universal Tractor, Moline Plow Co., Moline, IL.

Bottom: 1918-early 1920s, Indiana All Round Tractor, Indiana Silo & Tractor Co., Anderson, IN.

Early Tractors:

Top: 1919-(?) Bailor Motor Cultivator, Bailor Plow Mfg. Co., Atchison, KS.

Bottom: 1919 J.I. Case 12 h.p. Two-row Motor Cultivator, J.I. Case Plow Works, Racine, WI.

Early Tractors:

British tractors:

Top & Bottom: Two views of 1919-26 Austin 25 h.p. Tractor, Austin Motor Co.,
Longbridge, Birmingham.

Middle Left: 1919-24 Glasgow Three-Wheel Tractor,
D.L. Motor Mfg. Co., Motherwell.

Middle Right: 1921 Garner 30 h.p. Tractor, Henry Garner Ltd.,
Moseley Motor Works, Birmingham.

Early Tractors:

Crawler tractors:

Top Left: 1919-25 Holt Caterpillar Ten-Ton 40-60, Holt Mfg. Co., Stockton, CA.

Top Right: 1920-21 Stockton Model B Sure-Grip, Stockton Tractor Co., Stockton, CA.

Upper Middle Right: 1915-early 1920s Bean Track Pull 6-10, Bean Spray Pump Co., San Jose, CA.

Middle Left: 1920-24 Cletrac Model F 9-16, Cleveland Tractor Co., Cleveland, OH.

Lower Middle Right: 1916-19 Holt 25-45.

Bottom: 1922 Yuba Model 15-25, Yuba Mfg. Co., Maryville, CA.

Early Tractors:

Crawler tractors:

Top Left: 1920 Bates Steel Mule Model D 15-22,
Bates Machine & Tractor Co., Joliet, IL.

Top Right: 1919-25 Holt Caterpillar Five-Ton, Holt Mfg. Co., Stockton, CA.

Middle Left: 1922-25 Holt Caterpillar T-35 Two-Ton.

Middle Center; 1919-25 Cletrac Model W 12-20, Cleveland Tractor Co., Cleveland, OH.

Middle Right: 1917-19 Trundaar 20-35, Buckeye Mfg. Co., Anderson, IN.

Bottom: 1919-25 Cletrac Model W 12-20.

Early Tractors:

Top: 1919-30 Allis-Chalmers 18-30 Tractor, Allis Chalmers Mfg. Co., Milwaukee, WI.
It was later re-rated 20-35.

Middle Left: 1918 Little Giant Tractor, Little Giant Co., Mankato MN. At this time the
company built two models very similar in appearance, the Model A 26-35 and the Model B 12-22.

Bottom: 1921-27 Allis-Chalmers 12-20 Tractor.
It was later rerated 15-25 and became known as the Model L.

Early Tractors:

Top: 1918-22 Whitney 9-18 Tractor, Whitney Tractor Co., Cleveland, OH.

Middle Center: 1920 G-O Tractor, General Ordnance Co., Cedar Rapids, IA.

Middle Right: 1918 Gilson Guelph Tractor, Gilson Mfg. Co., Guelph, Ontario.
Gilson began manufacturing tractors at this time, and built three models,
the 11-20, 12-25, and 15-30. It gave up tractor production in 1922
after building only about a hundred machines.

Bottom: 1921-29 McCormick-Deering 15-30 Tractor,
International Harvester Co, Chicago.

Early Tractors:

Crawler Tractors:

Top: 1919-25 Best 60 Tracklayer, C.L. Best Gas Traction Co., San Leandro, CA.

Middle Left: 1919 Monarch 18-30 Neverslip, Monarch Tractor Co., Watertown, WI.

Middle Right: 1917-19 Cletrac Model H, Cleveland Tractor Co., Cleveland, OH.

Bottom: 1920-25 Best 30 Tracklayer.

Early Tractors:

Top: 1923-27 Wallis Model OK 15-27 Orchard Tractor, J.I. Case Plow Works, Racine. WI.

Middle Left: 1921-22 Hart-Parr "20," or Type B 10-20 Tractor, Hart-Parr Co., Charles City. IA.

Middle Right: 1924-26 Hart-Parr Type E 16-30 Tractor.

Bottom: 1922-27 Wallis Model OK 15-27 Tractor.

Early Tractors:

Top: 1923-24 version of McCormick-Deering 10-20 Tractor,
International Harvester Co., Chicago, IL.

Bottom: 1919-27 Twin City 12-20 Tractor, Minneapolis Steel & Machinery Co.,
Minneapolis, MN.

Early Tractors:

Rumely Oil Pull tractors, Advance-Rumely Thresher Co., La Porte, IN:

Top: 1924-27 Type L 15-25.

Bottom: 1918-24 Type K 12-20, The Type K was the lightest model in Rumely's "Heavyweight" lineup of 1918. The Type L was its replacement in the "Lightweight" series of 1924.

Early Tractors:

Top: 1918-28 Fordson Model F 9-18 Tractor, Ford Motor Co., Detroit, MI.
Fenders, which became optional on the Model F about 1923,
weren't shown in advertisements until about 1925.

Bottom: 1923-26 version of John Deere Model D 15-27 Tractor, Deere & Co., Moline, IL.

Early Tractors:

McCormick-Deering tractors, International Harvester Co., Chicago:

Top: 1924-32 Farmall, the revolutionary row-crop design.

Bottom; 1925-(?) version of 10-20.

Early Tractors:

Caterpillar crawler tractors, Caterpillar Tractor Co., Peoria, IL:

Top: 1925-28 Two-Ton. This updated version of the Holt Two-Ton model began production before the 1925 merger with C.L. Best.

Bottom: 1928-33 "Ten."

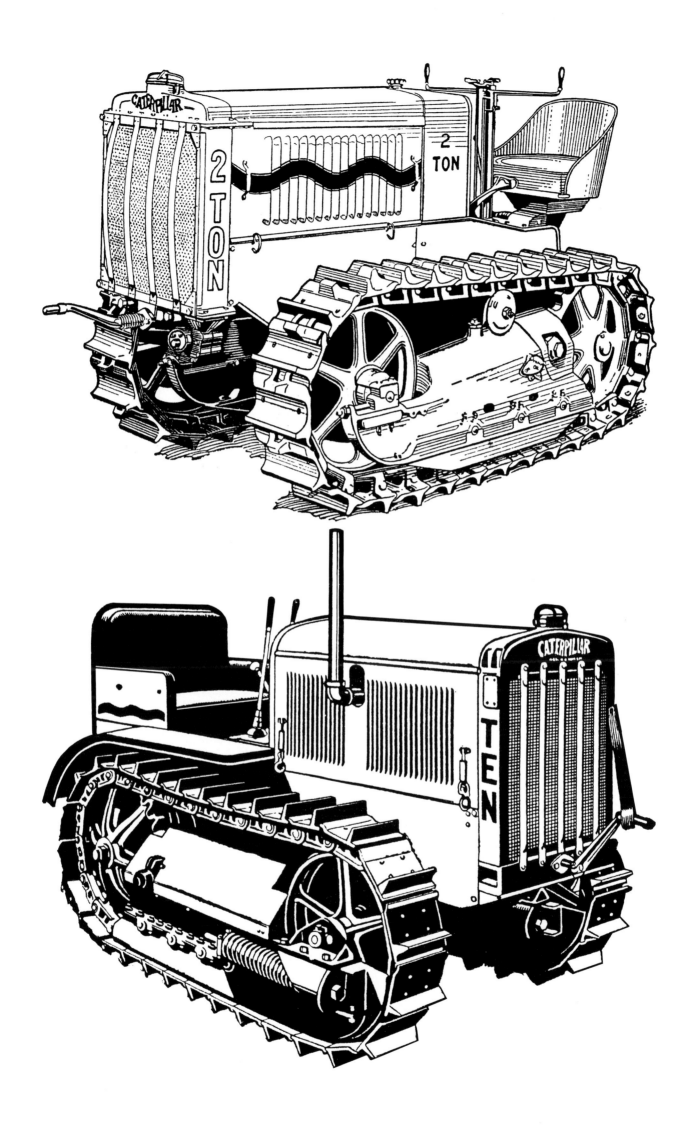

Early Tractors:

Rumely tractors, Advance-Rumely Thresher Co., La Porte, IN:

Top: A model from the 1928-30 "Lightweight" series, perhaps early version of the Type W 25-40.

Bottom: 1928-31 DoAll 20-hp. This model, which could be converted for
row-crop use, was developed from the Toro Motor Cultivator, the rights to which
Rumely purchased in 1927.

Early Tractors:

Top: 1927-37 Rock Island Model F 18-35 Tractor, Rock Island Plow Co., Rock Island, IL.

Bottom: 1927-31 Wallis 20-30 Orchard Tractor, J.I. Case Plow Works, Racine, WI.

Early Tractors:

Top: First advertised in late 1929, John Deere's GP Wide-Tread Tractor was the company's first row-crop model, allowing it to compete with the McCormick-Deering Farmall. It was built through 1935, Deere & Co., Moline, IL.

Bottom: 1929-40 Case Model L 26-40 Tractor, J.I. Case Co., Racine, WI.

Bibliography

Appleyard, John, *The Farm Tractor*, David & Charles, North Pomfret, VT, 1987.

Gay, Larry, "Why Did Charles Hart Leave His Company?" *Farm Collector*, Ogden Publications, Topeka, KS, May, 2007.

Gray, R.B., *The Agriculture Tractor: 1855-1950*, American Society of Agriculture Engineers, Saint Joseph, Michigan, 1975.

Johnson, Paul C., *Farm Power in the Making of America*, Wallace-Homestead Book Co., Des Moines, IA, 1978.

McKinley, Marvin, *Wheels of Farm Progress*, American Society of Agricultural Engineers, St. Joseph, Michigan, 1980.

Moore, Sam, "One Wicked Tractor," *Farm Collector*, Ogden Publications, Topeka, KS, April, 2007.

Moorhouse, Robert, *The Illustrated History of Tractors*, Chartwell Books, Edison, NJ, 1996.

Pripps, Robert N. & Morland, Andrew, *The Field Guide to Vintage Farm Tractors*, Voyageur Press, Stillwater, MN, 1999.

Sanders, Ralph W., *Vintage Farm Tractors: The Ultimate Tribute to Classic Tractors*, Town Square Books, Stillwater, MN, 1996.

Sanders, Ralph W., *Vintage International Harvester Tractors*, Town Square Books, Stillwater, MN, 1997.

Wendel, C.H., *Classic American Tractors: Oliver, Hart-Parr*, KP Books, Iola, WI, 2005.

Wendel, C.H., *Encyclopedia of American Farm Tractors*, Crestline Publishing, Sarasota, FL, 1979.

Wendel, C.H., *Standard Catalog of Farm Tractors: 1890-1980*, KP Books, Iola, WI, 2005.

Wik, Reynold M., *Steam Power on the American Farm*, University of Pennsylvania Press, Philadelphia, 1953.

Williams, Michael, *British Tractors for World Farming*, Blandford Press, Poole, Dorset, 1980.

Williams, Robert C., *Fordson, Farmall, and Poppin' Johnny: A History of the Farm Tractor and Its Impact on America*, University of Illinois Press, Urbana and Chicago, 1987.

Breeder's Gazette, Sanders Publishing Co., Chicago, IL

California Citrograph, Los Angeles, CA

California Cultivator, Los Angeles, CA

Country Gentleman, Curtis Publishing Co., Philadelphia, PA

Die Abendschule, St. Louis, MO

Farm & Ranch, Dallas, TX

Farm Implement News, Chicago, IL

Implement & Vehicle Journal, Dallas, TX

Mark Lane Express Agricultural Journal, London, England

New England Homestead, Springfield, MA

Ohio Farmer, Cleveland, OH

Pacific Rural Press, Los Angeles and San Francisco, CA

Progressive Farmer, Dallas, TX

Wallaces' Farmer, Des Moines, IA

Index

Pages numbers are cited in Roman type,
Plates numbers are cited in Italic type.

About the Author

BORN ON Oct. 2, 1941, in Lubbock, Texas, Jim Harter was largely self-taught as an artist. From 1969-72 he played a small part in creating posters for Austin's legendary rock venues, the Vulcan Gas Company and Armadillo World Headquarters. Influenced by San Francisco collage artist Wilfried Satty, Harter turned to making surrealist collages from 19th century engravings. In 1976 he moved to New York, becoming a freelance illustrator, and editor of clip-art books for Dover, and later for other publications. Since then two books of his collages, *Journeys in the Mythic Sea* and *Initiations in the Abyss* have been published, as well as two railroad history books illustrated entirely with Victorian engravings. In 1984, he began painting using an old-master technique which he learned from Carlos Madrid. This work owes inspiration to Symbolism, Surrealism, Fantastic Realism, and an exposure to Eastern Philosophy. During the early 1980s Harter became friends with Dr. Jean Letschert, a Belgian visionary painter and former student of Rene Magritte. He also met members of Holland's Metarealist group, and fantastic realist painters in New York. In 1986 Harter moved to San Antonio, Texas where he remains today. In the last two years he has returned to his collage work, digitally colorizing a number of his creations.

Published Works

Scientific Instruments and Apparatus: CD-ROM and Book, Dover Publications, Mineola, NY, 2007.

Insects: CD-ROM and Book, Dover Publications, Mineola, NY, 2007.

World Railways of the Nineteenth Century: A Pictorial History in Victorian Engravings, The Johns Hopkins University Press, Baltimore, 2005.

Nautical Illustrations: 681 Permission-free Illustrations from Nineteenth Century Sources, Dover Publications, Mineola, NY, 2003.

Initiations in the Abyss: A Surrealist Apocalypse, Wings Press, San Antonio, 2003.

Landscapes and Cityscapes for Artists and Craftspeople: From 19th-Century Sources, Dover Publications, Mineola, NY, 1999.

American Railroads of the Nineteenth Century: A Pictorial History in Victorian Wood Engravings, Texas Tech University Press, Lubbock, 1998.

The Ultimate Angel Book, Gramercy Books, New York, 1995.

Images of Medicine, Bonanza Books, New York, 1991.

Images of World Architecture, Bonanza Books, New York, 1990.

The Plant Kingdom Compendium, Bonanza Books, New York, 1988.

Journeys in the Mythic Sea: An Innerspace Odyssey, Harmony Books, New York, 1985.

Hands: A Pictorial Archive from Nineteenth Century Sources, Dover Publications, New York, 1985.

Transportation: A Pictorial Archive from Nineteenth Century Sources, Dover Publications, New York, 1984.

Music: A Pictorial Archive of Woodcuts and Engravings, Dover Publications, New York, 1980.

Men: A Pictorial Archive from Nineteenth Century Sources, Dover Publications, New York, 1980.

Animals: A Pictorial Archive from Nineteenth Century Sources, Dover Publications, New York, 1980.

Food and Drink: A Pictorial Archive from Nineteenth Century Sources, Dover Publications, New York, 1979.

Women: A Pictorial Archive from Nineteenth Century Sources, Dover Publications, New York, 1978.

Harter's Picture Archive for Collage and Illustrations, Dover Publications, New York, 1978.

Die Gretchen, Speleo Press, Austin, 1973.

Wings Press was founded in 1975 by Joanie Whitebird and Joseph Lomax, both deceased, as "an informal association of artists and cultural mythologists dedicated to the preservation of the literature of the nation of Texas." Publisher, editor and designer since 1995, Bryce Milligan is honored to carry on and expand that mission to include the finest in American writing—meaning all of the Americas—without commercial considerations clouding the choice to publish or not to publish.

Wings Press publishes multicultural books, chapbooks, ebooks, CDs and DVDs that, we hope, enlighten the human spirit and enliven the mind. Every person ever associated with Wings has been or is a writer, and we know well that writing is a transformational art form capable of changing the world, primarily by allowing us to glimpse something of each other's souls. Good writing is innovative, insightful, and interesting. But most of all it is honest.

Likewise, Wings Press is committed to treating the planet itself as a partner. Thus the press uses as much recycled material as possible, from the paper on which the books are printed to the boxes in which they are shipped. All inks and papers used meet or exceed United States health and safety requirements.

As Robert Dana wrote in *Against the Grain,* "Small press publishing is personal publishing. In essence, it's a matter of personal vision, personal taste and courage, and personal friendships." Welcome to the Wings Press community of readers.

Colophon

This first edition of *Early Farm Tractors: A History in Advertising Line Art,* by Jim Harter, has been printed on 70 pound Edwards Brothers Matte Coated Paper containing a percentage of recycled fiber. Titles have been set in Colonna MT type, the text in Adobe Caslon type. All Wings Press books are designed and produced by Bryce Milligan.

On-line catalogue and ordering available at
www.wingspress.com

Wings Press titles are distributed to the trade by the
Independent Publishers Group
www.ipgbook.com

Also available as an ebook.